574.88076
F88p

139689

DATE DUE			

PROBLEMS FOR MOLECULAR BIOLOGY

PROBLEMS FOR MOLECULAR BIOLOGY

With Answers and Solutions

David Freifelder

UNIVERSITY OF CALIFORNIA, LA JOLLA
UNIVERSITY OF ALABAMA

JONES AND BARTLETT PUBLISHERS, INC.
BOSTON   PORTOLA VALLEY

Editorial offices: Jones and Bartlett Publishers, Inc., 30 Granada Court, Portola Valley, CA 94025

Sales and customer service offices: Jones and Bartlett Publishers, Inc., 20 Park Plaza, Boston, MA 02116

Library of Congress Cataloging in Publication Data

Freifelder, David Michael, 1935–
  Problems for Molecular biology.

  Includes index.
  1. Molecular biology—Problems, exercises, etc. I. Freifelder, David Michael, 1935– . Molecular biology. II. Title.
QH506.F734 1983**b**   574.8'8'076   83-13576
ISBN 0-86720-013-8 (pbk.)

ISBN 0-86720-013-8

Publisher Arthur C. Bartlett
Book and cover design Hal Lockwood
Production Bookman Productions
Printer and binder Banta Company

Printed in the United States of America
Printing number (last digit)   10 9 8 7 6 5 4 3

Contents

Preface

In any experimental science it is important that students have practice in dealing with experimental observations. Carefully selected problems and questions can give such practice; furthermore, they provide students with an opportunity to test his or her understanding of textbook and lecture material and to put together various facts in order to draw conclusions not explicitly stated elsewhere. This book consists of such questions and problems.

The organization of this book follows that of the accompanying text—*Molecular Biology, A Comprehensive Introduction to Prokaryotes and Eukaryotes*—also published by Science Books International. However, because of the large number of questions on a variety of subjects, the book should also be of value to students using other molecular biology texts, as well as in courses in biochemistry, genetics, cell biology, and advanced molecular biology.

This book is meant to be a study guide for the student rather than a source of homework problems for the instructor. For this reason, the material ranges in difficulty considerably and covers a large number of topics. Of prime importance are those questions marked with a solid circle. These are elementary (though not necessarily simple) questions and are designed to ensure that the student understands various terms and knows basic facts and concepts. It is important that the student be able to answer these questions before proceeding to more complex material or to later chapters in the text.

If answering the questions is to be a learning experience, answers must be provided. A single word or number is insufficient to tell a student who has answered a problem incorrectly where he or she went wrong. Therefore, except for the most elementary questions, answers are given in some detail and calculations are given for those problems requiring numerical answers.

Writing problems is a time-consuming operation compounded by the worry that a problem might be thought ambiguous or be misconstrued by the student who perceives details more sharply than concepts or, even worse, might not be answerable. All teachers face this problem but, after many years, usually accumulate a set of unambigous, solvable problems. Since the point of view of different instructors may vary, I have considered it necessary when writing this book to enlist the aid of other molecular biologists to supplement my own collection of roughly 600 problems. I would like to express thanks to Rich Calendar, Ellen Daniell, Hatch Echols, Peter Geiduschek, Julie Marmur, Frank Stahl, and Bob Warner for supplying problems. A judgment of the clarity and lack of ambiguity of a problem can only be made by students. Thus, I have enlisted the aid of many students in reading the problems and checking the answers. Two groups have been used—undergraduates in the Department of Biochemistry at Brandeis University, who took my course in 1980, and a group of graduate students and postdoctoral fellows in the Department of Molecular Biology at the University of California, Berkeley. I owe all of these people great thanks.

San Diego, California   David Freifelder
October, 1982

1

Review of Biology and Genetics

● 1-1 Is an experiment in which a radioactive DNA precursor is added to a bacterial culture in order to study DNA synthesis an (a) *in vitro* or an (b) *in vivo* experiment?

1-2 Suggest several reasons why the product of an enzymatic reaction might not be found in a crude extract.

● 1-3 (a) Is a cell in which the DNA is enclosed in a nuclear membrane a prokaryote or a eukaryote?

(b) Classify the following cells as either prokaryotes or eukaryotes: bacteria, fungi, algae, yeasts, amoebae, wheat cells, human liver cells.

1-4 A bacterium divides every 35 minutes. If a culture containing 10^5 cells per ml is grown for 175 minutes, what will be the cell concentration?

● 1-5 Match the terms with the definitions:

Terms
1. Minimal medium. 2. Prototroph. 3. Plating. 4. Auxotroph.

Definitions

A. A cell that can grow in a minimal medium.

B. Depositing bacteria or phage on an agar surface for growth.

C. A cell that requires for growth organic substances other than an organic carbon source.

D. A growth medium containing salts and only one organic compound, which is used as a carbon source.

1-6   What is the term used to describe the growth of a bacterial culture in which the doubling time is constant?

1-7   Suggest reasons for bacteria growing in liquid medium to enter stationary phase.

1-8   What is the cell density of a bacterial culture if plating 0.1 ml of a 10^5-fold dilution of the culture yields 68 colonies.

●1-9   Define (a) lysozyme and (b) lysis.

●1-10   Define haploid and diploid.

1-11   Distinguish a primary culture from an established cell line.

●1-12   (a) Does the symbol *lac*$^+$ refer to the genotype or the phenotype of a cell?

(b) What notation would indicate that a cell is resistant to the antibiotic penicillin?

(c) If a cell has three genes involved in synthesis of proline, namely, *proA*, *proB*, and *proC*, would there be any difference in the genotype of a cell designated *proA*$^+$*proB*$^+$*proC*$^+$ and one designated *pro*$^+$?

(d) Referring to part (c), how would you write the phenotype of a cell whose genotype is *proA*$^+$*proB*$^-$*proC*$^+$?

●1-13   Define absolute defective mutant, conditional mutant, and temperature-sensitive mutant.

1-14   The frequency of producing mutant *arb*$^-$ is 2 x 10^{-6} per generation and mutant *snd*$^-$ is 8 x 10^{-5}. What is the frequency of producing an *arb*$^-$*snd*$^-$ mutant in a single event?

1-15   A colleague suggests to you that glucose is a substance that freely diffuses through the *E. coli* cell wall. You have in your refrigerator an *E. coli* mutant with the following property. If glucose is the sole carbon source in the growth medium, it grows normally at 34°C but cannot grow at 42°C; with any other carbon source it grows well at both temperatures.

(a) Can you conclude that there is a glucose transport system?

(b) Suppose, by biochemical techniques, you make the following observation. Glucose is not metabolized in this mutant at 42°C; however, if the cell culture is treated with the enzyme lysozyme, which removes a portion of the cell wall, glucose metabolism can occur at 42°C. Does this change your conclusion?

1-16   Mutants that fail to synthesize a substance X have been found in four complementation groups, none of which are cis-acting. How many proteins are required to synthesize X?

1-17   Four genes kyuA, kyuB, kyuC, and kyuQ are known to be required to synthesize substance Q and each biochemical reaction can be detected. The reaction sequence is P → B → C → A → Q in which the product of a gene kyuX is needed to synthesize substance X. Addition of ^{14}C-P yields ^{14}C-Q.

(a) A mutant is found for which addition of ^{14}C-P yields ^{14}C-A but no ^{14}C-Q. In what gene is this mutation?

(b) Another mutant is found for which there is no conversion of ^{14}C-P to any other substance. Furthermore, addition of ^{14}C-A fails to yield ^{14}C-Q. What kind of mutant is this one?

1-18   Study of a temperature-sensitive mutant in the gene memA indicates that there is no active MemA enzyme at 42°C, though at 34°C active enzyme is present. The enzyme is purified from cells grown at 34°C and its activity is tested at 42°C. It remains active at this temperature. What conclusion can be drawn about the relation between the memA gene and the MemA enzyme?

1-19   In E. coli phage β some mutations in the phage gene G are compensated for by an additional mutation in gene H and some mutations in gene H are compensated for by mutations in gene G. What do these facts indicate?

1-20   Approximately 10^8 E. coli cells of an auxotrophic strain are plated on complete medium and form a bacterial lawn. Replica plates are prepared containing minimal medium supplemented by leucine, thymine, and proline, as shown in Figure 1-20 (page 3). (a) What is the genotype of the strain? (Note that there is not a lawn on any of the supplemented plates.)

(b) Explain the colonies that appear on the replica plates.

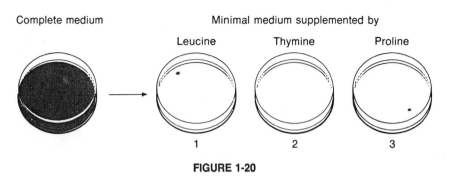

Complete medium   Minimal medium supplemented by

Leucine   Thymine   Proline

1   2   3

FIGURE 1-20

1-21 An Hfr cell with the genotype *arg⁺leu⁺pro⁺thr⁺thy⁺ser⁺-his⁺lac⁻str-s* is mated with an *F⁻* cell which has the genotype *arg⁻leu⁻pro⁻thr⁻thy⁻ser⁻his⁻lac⁻str-r*. Lac⁺Str-r recombinants are selected by plating on agar containing all seven nutrients, lactose, and streptomycin. A velvet pad is pressed on this master plate, thereby picking up the colonies, and then pressed on the six plates shown in Figure 1-21, with the indicated supplements. Colonies grew as shown. What are the genotypes of the numbered colonies?

1-22 Often dyes are incorporated into agar in order to determine whether a bacterium can utilize a particular sugar as a carbon source. For instance, on EMB agar containing a sugar X, suppose that a bacterium having the phenotype X⁺ makes a purple colony and another having the phenotype X⁻ makes a pink colony. (Fermentation of the sugar produces acid which changes the dye to a purple color.) If the EMB agar contained rhamnose, xylose, and galactose, what would be the color of the colonies produced by bacteria having the following phenotypes? Rha⁺Xyl⁺Gal⁺, Rha⁻Xyl⁻Gal⁺, Rha⁺Xyl⁻Gal⁺, Rha⁻Xyl⁺Gal⁻, Rha⁻Xyl⁻Gal⁻.

1-23 On EMB agar Lac⁺ and Lac⁻ colonies are purple and pink respectively. A bacterial strain, S, is found which produces almost 99 percent purple colonies and 1 percent pink colonies when grown at 30°C; at 42°C almost all colonies are pink. If S bacteria are grown at 42°C for several generations and then plated at 30°C, half of the colonies formed are purple and half are pink. If a pink colony from a 42°C plate is resuspended and replated at 30°C, only pink colonies result; when purple colonies from a 30°C plate are replated, most colonies are purple at 30°C and all are pink at 42°C. When strain S is plated at 42°C, about 1 colony in 10⁴ is purple. Such purple colonies fall into three

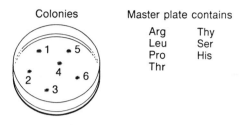

Colonies

Master plate contains

Arg   Thy
Leu   Ser
Pro   His
Thr

Replica plates contain amino acids, as indicated

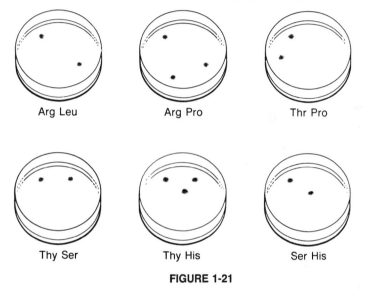

Arg Leu   Arg Pro   Thr Pro

Thy Ser   Thy His   Ser His

FIGURE 1-21

classes: (1) those which remain purple at 42°C but still produce about 1 percent pink colonies at 30°C or 42°C; (2) those which are purple at 42°C but never (less than 1 in 10^5) produce pink colonies at 42°C or 30°C; and (3) those which are like class 2 but have the additional property of being resistant to phage T1 (the original strain was T1-sensitive). Other classes are possible but will not be considered here. What is the relevant genotype of the original strain S? What are the genotypes of the three classes of variants? (Note: T1-sensitivity is dominant over T1-resistance.)

1-24 One cell of *E. coli* contains about 10^{-14} g of DNA. A DNA strand is 20 angstroms wide and has a mass of about 2×10^6 daltons for each micrometer of length. What fraction of the volume of *E. coli* is DNA?

1-25 *E. coli* has a cylindrical shape about 1 µm in diameter and 3 µm long. The doubling time of *E. coli* when growing on nutrient agar is about 25 minutes. After 12 hours of growth, a colony is roughly 2 mm in diameter and 1/2 mm high. Have all of the cells been growing for 12 hours?

1-26 Hair color in some animals can be black *(BB)*, gray *(Bb)*, or white *(bb)*. If a black and gray animal mate and produce one offspring, what is the probability that it is gray? If there are two offspring, what is the probability that both are gray? If there are three offspring, what is the probability that only one is gray?

•1-27 Two parents each with genotype *AaBb* mate and produce 16 offspring. How many of the offspring would you expect to be homozygous recessive for both genes?

1-28 An animal can have red (*RR* or *Rr*) or blue (*rr*) eyes. If they are also *tt*, their eyes are colorless (*TT* and *Tt* give color). If *RrTt* mates with *RRtt*, what is the probability of getting a blue-eyed individual?

1-29 An animal has a single gene for tail shape. If a fat-tailed animal is mated to another fat-tailed one, only fat-tailed animals result. If a fat one and a thin one are mated, half of the progeny are thin. If two thin animals mate, there are always, on the average, twice as many thin progeny as fat progeny. Identify the genotypes corresponding to the fat and thin phenotypes. Hint: Determine the allele that is dominant.

1-30 Color blindness in humans is inherited as a sex-linked recessive trait. Write as far as possible the genotype of each person represented in the pedigree shown in Figure 1-30. Black = colorblind.

1-31 Following is a list of mutational changes. Which would be recessive in a heterozygote and which would be dominant? (a) The mutant protein has no activity but the total number of molecules made by the gene is in excess of that needed for normal biological function.
(b) A protein contains four subunits. Both mutant and good protein units can interact. One defective subunit eliminates activity.

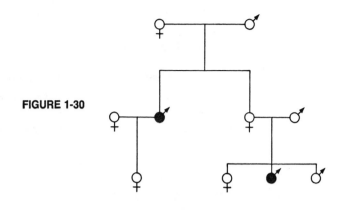

FIGURE 1-30

(c) A mutant enzyme fails to carry out a particular chemical reaction.

(d) A mutant enzyme reverses the chemical reaction of a normal enzyme.

● 1-3 2   (a) Suppose the following recombination frequencies occur between the indicated markers: a x c , 2 percent; b x c , 13 percent; b x d, 4 percent, a x b, 15 percent, c x d, 17 percent; a x d, 19 percent. What is the gene order?
(b) In the cross aBd x AbD, what is the frequency of getting ABD progeny?

1-3 3   A linear phage with genotype Ab a is crossed with another phage with genotype aBD. The following recombinants are observed at the indicated frequencies: abd, 2 percent; ABd, 3 percent; aBd, 0.06 percent. What is the gene order?

1-3 4   In the cross $abcd$ x $ABCD$, the following recombinants were found at the indicated frequencies: $ABCd$, 3 percent; $abcD$, 3 percent; $AbcD$, 0.03 percent; $AbCd$, 0.0006 percent; $ABcD$, 0.06 percent. What is the gene order?

1-3 5   Four genes have the order a b d e. The recombination frequencies between various pairs are: a x b, 1 percent; b x d, 2 percent; d x e, 3 percent.
(a) What is the frequency of production of AE recombinants in the cross $AbDe$ x $aBdE$?
(b) What fraction of the AE recombinants will be $ABDE$?

1-3 6   In a cross between two phage having genotypes EFG and efg, 1000 progeny were analyzed. The number of phage having each of the eight possible genotypes were as follows: *efg*, 396; *EFG*, 406; *eFg*, 23; *efG*, 1; *EfG*, 25; *Efg*, 75; *eFG*, 73; *EFg*, 1. Construct a map showing the positions of the genetic markers.

1-37 An animal has a coat that is either red or white. Its tail is either long or short. When (red, long) is mated with (red, long), all progeny have red coats. When (white, long) is mated with (white, long), both red- and white-coated progeny result. When long is mated with long, only long-tailed progeny arise. When (red, short) mates with another (red, short), both short- and long-tailed progeny arise.
(a) What are the genotypes corresponding to the red, white, long, and short phenotypes?
(b) When (white, short) mates with another (white, short), the resulting progeny have these frequencies: 1/9 red, long; 2/9 red, short; 2/9 white, long; and 4/9 white, short. Explain how these frequencies arise.

1-38 A pterodactyl can have blue or white eyes, and long or short wings. Blue eyes and long tails are dominant characteristics. A blue-eyed, long-winged male that is known to be heterozygous for both loci mates with a blue-eyed, long-winged female that is also heterozygous. The following phenotypes with the indicated frequencies are found among the progeny: 3/8 blue, long female; 1/8 blue, short female; 3/16 blue, long male; 1/16 white, short male.
(a) What gene is carried on the X chromosome (assuming that sex is determined as in humans?)
(b) An archaeopteryx can also be blue- or white-eyed and long- or short-winged. Again, a heterozygous blue-eyed, long-winged male mates with a blue-eyed, long-winged female. This time the phenotypic frequencies among the progeny are the following: 1/2 blue, long female; 1/4 blue, long male; 1/4 white, short male. Another heterozygous blue eyed, long-winged female mates with the same male and there are 1/2 blue, long female; 1/4 blue, short male; 1/4 white, long male. How does an archaeopteryx differ from a pterodactyl with respect to eye color and wing length?

1-39 Suppose you are given a mutant strain of *E. coli*. In checking its genotype you test growth requirements at two temperatures. You plate 10^8 cells at either 25°C or 42°C on minimal agar plates supplemented with various amino acids. The amount of growth is indicated in Table 1-39A.

TABLE 1-39A

Supplement	Temperature (°C)	Growth of colonies
His, Trp	25	None
His, Leu	25	10
Leu, Trp	25	Confluent
His, Leu, Trp	25	Confluent
His, Trp	42	None
His, Leu	42	8
Leu, Trp	42	12
His, Leu, Trp	42	Confluent

(a) What is the genotype of the strain?

(b) Why is there confluent growth on the plate containing leucine and tryptophan at 25°C but not at 42°C?

(c) Why are there no colonies on the plates containing histidine and tryptophan?

(d) What is the expected genotype of the colonies that grow at 42°C on the agar containing histidine and leucine?

(e) What is the expected genotype of the colonies that grow at 42°C on the agar containing leucine and tryptophan?

(f) Suppose you have an Hfr strain with genotype met^--$his^+leu^+trp^+$. This Hfr is known to transfer met very late. You mix 10^8 bacteria of the Hfr strain and 10^9 of your mutant bacteria and allow mating to occur for several hours. You plate a 10,000-fold dilution of the mixture on the following plates at 42°C and incubate them for two days. You observe the results shown in Table 1-39B.

TABLE 1-39B

Nutrient or nutrients in agar	Number of colonies found
Trp, His	250
Leu, His	50
Leu, Trp	500
His	10

(1) Which genes entered first, second, and third?

(2) You now know the relative order of these three markers, but

you know nothing about their exact location on the chromosome. Describe an experiment that will tell you where these markers are located on the chromosome.

(3) What is the purpose of the *met⁻* mutation in the Hfr strain in this experiment?

1-40   A DNA fragment was obtained from a bacterial population whose genotype is *pur⁺pro⁻his⁻*. The fragment was cut from the chromosome as a linear piece of DNA of fixed size . Cuts were made at random so the fragment might have contained one, two, or three of the genes. A mixture of these fragments was added to a culture of a recipient bacterium (which could take in DNA) having the genotype *pur⁻pro⁺his⁺*, and *pur⁺* recombinants were selected. The *pur⁺* recombinants were tested for the unselected *pro* and *his* markers, and the data in Table 1-40 were found. What is the gene order and what is the relative distance between the mutations?

TABLE 1-40

Genotype	Number of colonies
pro⁺ his⁺	103
pro⁻ his⁺	24
pro⁺ his⁻	158
pro⁻ his⁻	1

1-41   An Hfr whose genotype is *a⁺b⁺c⁺d⁺str-s* is mated with an *F⁻* strain whose genotype is *a⁻b⁻c⁻d⁻str-r*. Mating is interrupted at various times and each sample that is obtained is plated on four different agar types on which *a⁺str-r*, *b⁺str-r*, *c⁺str-r* recombinants can grow. A plot of the number of recombinants of each type as a function of time generates a set of time-of entry curves, so called because the intersection of each line with the time axis defines the time at which a particular allele first enters an *F⁻* cell. Consider the time-of-entry curves shown in Figure 1-41. Which of the following statements (1 or 2) is the explanation for the fact that the curves form plateau regions at lower values as one proceeds rightward on the graph?

(1) The probability of a gene being inserted into the *F⁻* chromosome decreases with time.

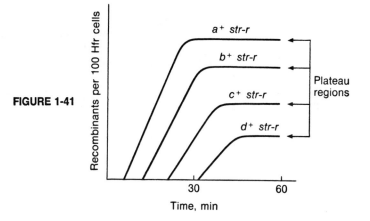

FIGURE 1-41

(2) The probability of a gene being transferred to the F^- cell decreases with time.

1-42 Consult the genetic map of *E. coli* in Appendix A. Suppose an F plasmid were inserted into the *strA* gene of *E. coli*. If the resulting Hfr strain was then shown to transfer the *fda* gene within a few minutes after the onset of conjugation, at what time after conjugation started would this strain transfer the *argC* gene?

1-43 An Hfr donor of genotype $a^+b^+c^+d^+str$-s is mated with an F^- recipient having genotype $a^-b^-c^-d^-str$-r. Genes *a*, *b*, *c*, and *d* are spaced equally. A time-of-entry experiment is carried out and the data shown in Table 1-43 are obtained. What are the times of entry for each gene? Explain the low recombination frequency in the plateau region for d^+str-r recombinants..

1-44 The order of four genes in an Hfr strain is *a b c d*. In a cross between an Hfr donor that has genotype $a+b+c+d+x^-$ str-s and a female that has genotype $a^-b^-c^-d^-x^+str$-r, 90 percent of the d^+str-r recombinants are x^- and 100 percent of the c^+d^+str-r recombinants are x^-. The times of entry of *a*, *b*, *c*, and *d* are 5, 10, 15, and 20 minutes; the *str* gene enters at 55 minutes. Where is *x* located?

1-45 An Hfr donor whose genotype is $a^+b^+c^+str$-s mates with a female whose genotype is $a^-b^-c^-str$-r; the order of transfer is *a b c*. None of these genes are transferred early, the distance between *a*

TABLE 1-43

Time of mating, in min	Number of recombinants of indicated genotype per 1000 Hfr			
	a^+str-r	b^+str-r	c^+str-r	d^+str-r
0	0.01	0.006	0.008	0.0001
10	5	0.1	0.01	0.0004
15	50	3	0.1	0.001
20	100	35	2	0.001
25	105	80	20	0.1
30	110	82	43	0.2
40	105	80	40	0.3
50	105	80	40	0.4
60	105	81	42	0.4
70	103	80	41	0.4

and b is the same as the distance between b and c, and none of the markers are near str. Recombinants are selected as usual by plating on agar lacking particular nutrients and containing streptomycin. Which of the following are true (several answers are)? Explain.

(a) a^+str-r colonies > c^+str-r colonies.

(b) b^+str-r colonies < c^+str-r colonies.

(c) a^+b^+str-r colonies < b^+str-r colonies.

(d) a^+b^+str-r colonies = b^+str-r colonies.

(e) Most a^+c^+str-r colonies will also be b^+.

(f) Most b^-c^+str-r colonies will also be a^-.

(g) $a^+b^+c^-str$-r colonies < $a^+b^-c^-str$-r colonies.

1-46   An Hfr with genotype $a^+b^+c^+d^+str$-s is mated with a female with genotype $a^-b^-c^-d^+str$-r. At various times the culture is shaken violently to break apart pairs and the cells are plated on agar having the composition shown in Table 1-46A. (+ means that the nutrient is present.) The numbers of colonies growing on the agar are shown in Table 1-46B. One hundred colonies from the 25-min plates are picked and transferred to a dish containing agar of type 4, which contains A, B, C, and streptomycin, but no D. The number of colonies which grow on type 4 is shown in Table 1-46C.
(a) What is the order of the genes a, b, c, and d?
(b) Roughly how many colonies would be expected at various times on agar containing C and streptomycin but no A or B?

TABLE 1-46A

Agar of type	Str	A	B	C	D
1	+	+	+	−	+
2	+	−	+	+	+
3	+	+	−	+	+

TABLE 1-46B

Time of sampling, min	Number of colonies on agar of type		
	1	2	3
0	0	0	0
5	0	0	0
7.5	100	0	0
10	200	0	0
12.5	300	0	75
15	400	0	150
17.5	400	50	225
20	400	100	250
25	400	100	250

TABLE 1-46C

Colonies taken from agar of type	Number of colonies on agar of type 4
1	89
2	51
3	8

1-47   When an F^+ cell mates with an F^- cell, the recipient is converted
       to a male with very high efficiency. This is not so when the
       donor is an Hfr. How do you explain this?

1-48   After a brief mating between an Hfr whose genotype is pro^+pur^+-
       lac^+ and a female whose genotype is $F^-pro^-pur^-lac^-str$-r, many
       $lac^+pur^+pro^-str$-r recombinants are found. A few $pro^+lac^-pur^-$-
       str-r recombinants also arise, and all of these are Hfr donors.
       Explain this result and state the location of F in the Hfr. (The
       three genes pro, lac, and pur are very near one another.)

1-49   Suppose you have isolated two independent arginine-requiring
       (Arg$^-$) mutant strains from a parent $E.$ $coli$ strain which already
       requires methionine (Met$^-$) and is resistant to streptomycin
       (Str-r). You mate the two mutants (1 and 2) with an Hfr strain
       whose genotype is arg^+met^+str-s. Using the interrupted mating
       technique you obtain the time-of-entry curves shown in Figure
       1-49. Draw a genetic map which explains the difference observed
       in the two matings. (See problem 1-41 for a description of
       time-of-entry curves).

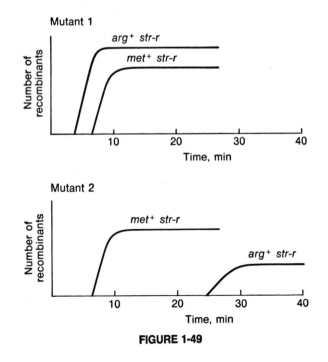

FIGURE 1-49

1-50   Suppose you collect a large number of galactose–requiring (Gal⁻) bacterial mutants and identify three closely-linked genes (designated *galA*, *galB*, and *galC*) by complementation and rough mapping studies. You wish to order these genes and learn something about the genetic structure of the "galactose region" of DNA. To order the genes, you mate an Hfr having the genotype *bio⁺ gal⁺ str-s* with various *F⁻* strains having the genotype *gal⁻bio⁻str-r*. The Hfr transfers the *bio* locus later than the *gal* locus. You select *bio⁺str-r* recombinants, measure the fraction of these that have the genotype *gal⁺*, and obtain the data shown in Table 1-50. What is the gene order relative to the *bio* locus?

TABLE 1-50

gal mutation	Number of gal⁺/number of bio⁺ in cross
galA⁻	0.65
galB⁻	0.72
galC⁻	0.84

1-51   You are interested in the biosynthetic pathway of compound X and isolate 10 different (independently isolated) mutants of *E. coli* which require compound X for growth. The mutations are mapped and their approximate positions are given in Figure 1-51.
(a) What is the minimum number of genes involved in the synthesis of compound X?
(b) Why must your answer be the minimum estimate?

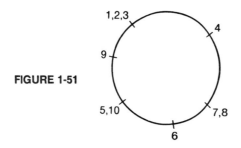

FIGURE 1-51

of glucose, (4) production of an amino acid by enzymatic hydrolysis of a protein.

2-7   Could a bacterium grow in the following growth medium? 0.02 M
       KH_2PO_4, 0.043 M Na_2HPO_4, 0.03 M $(NH_4)_2SO_4$, 0.0002 M $CaCl_2$,
       0.002 M $MgSO_4$, 0.01 M glucose?

2-8   Some bacteria that grow well in M9 medium (a minimal medium in
       which the carbon source is glucose) fail to grow if the glucose
       is replaced by sucrose (a glucose dimer). Explain.

2-9   Growth media are often prepared without the carbon source to
       avoid accidental contamination by bacteria during subsequent
       handling. Sterile glucose is added when needed. However, if one
       is not careful during handling of the glucose-free medium, a
       greenish sludge often accumulates at the bottom of the container.
       If an aliquot of the sludge is spread on a typical agar used to
       grow bacteria, no bacterial colonies are found after a 16-
       hour incubation at 37°C (a standard incubation time). What might
       the sludge be?

2-10   A bacterium which does not have a nutritional requirement for any
       known amino acid is placed in a minimal growth medium that
       contains the essential salts and the amino acid glycine but lacks
       any sugar. The bacterium grows and divides repeatedly but growth
       is at about half the rate that would occur if glucose were
       present. Explain this phenomenon.

2-11   What is the principal intracellular role of both the glycolysis
       reaction and the Krebs cycle?

2-12   How do cells store the energy generated during metabolic proc-
       esses usually so the energy can be available at a later time?

2-13   Explain why, when glucose is the sole carbon source, bacteria
       grow much more slowly (by about a factor of 10 to 20) in the
       absence of oxygen that in the presence of oxygen.

2-14   Bacteria can use a very large number of organic compounds (for
       example, sugars, alcohols, and amino acids) as a carbon source.
       Animal cells usually require a single sugar—they are basically
       glucose burners. Why might this be expected?

2-15 Some strains of bacteria excrete large amounts of nucleases. Propose a reason why such excretion might be useful to the bacteria in nature. Why do you think that the mammalian pancreas secretes large amounts of ribonucleases and deoxyribonucleases?

2-16 What is the essential difference between an enzyme and a coenzyme? What kinds of reactions are typically carried out by coenzymes?

2-17 In a preliminary study of the synthesis of a substance G, geeA⁻ mutants were found that make the cell auxotrophic for G. One of the mutants, geeB⁻, which is phenotypically Gee⁻, was also phenotypically Jay⁻, in that the mutant required both G and J for growth. A possible explanation is that the pathway for synthesis of G included J as a precursor.
(a) To test this hypothesis radioactive J was added to a culture of growing cells. If radioactive G were not produced, would this observation mean that J is not a precursor of G?
(b) Let us assume that J is not a precursor to G. How, then, might you explain the Jay⁻ phenotype of the geeB⁻ mutant?

2-18 If glucose labeled with ^{14}C in position 1 is added to a bacterial culture under anaerobic conditions, which carbon atom of lactic acid would be labeled?

2-19 An E. coli cell is cylindrical, about 1 μm in diameter and 3 μm long. At pH 7, how many H⁺ ions are there in a volume the size of one cell? Note that, statistically, this number would vary greatly from cell to cell. Since many reactions are strongly pH-dependent, do you think that such fluctuations, if they in fact occur, would introduce significant heterogenity among the individual cells of a population? Explain. E. coli is capable of normal growth in nutrient media having a wide range of pH values. This is probably because the pH within the cell is not the same as that of the surrounding medium. How do you think a cell might regulate its internal pH?

●2-20 In a biosynthetic reaction, must the complete reaction sequence, as it occurs in a cell, have a negative value of ΔG?

2-21 A DNA polymerase can make DNA from nucleotides and a nuclease can degrade DNA to nucleotides. Are these facts consistent with the statement that an enzyme only affects the rate and not the equilibrium of a reaction?

• 2-22   Explain in simple terms the principal role played by high-energy phosphate compounds in promoting chemical reactions.

2-23   Bacteria grow at different rates when different carbon sources are provided in a growth medium. What two features of the metabolism of carbon compounds principally determine the growth rate?

3

Macromolecules

3-1   Which amino acid is unable to form a proper peptide bond?

3-2   Many proteins bind metal ions quite tightly. Which amino acid would probably be responsible for this binding? If the ion were Hg^{2+}, which amino acids would most likely be involved?

3-3   What is a common state of cysteine in a protein?

3-4   Proteins strongly absorb ultraviolet light in a wavelength range of 275 to 290 nm. Which amino acids are responsible for this absorption?

3-5   Which amino acids can engage in hydrogen bonding?

3-6   A protein contains four cysteines. If all were engaged in disulfide bonds and if all possible pairs of cysteines could be joined, how many different protein structures would be possible?

3-7   Name the kinds of bonds in proteins in which each of the following amino acids might participate: cysteine, lysine, isoleucine, glutamic acid.

● 3-8   Consider the chemical structure of the heptapeptide
          N-Ala-Trp-Ser-Pro-Leu-Ile-Gly-COOH. How many peptide bonds
          are there?

● 3-9   What chemical groups are at the ends of a protein molecule?

3-10   Only one amino acid side chain engages in covalent bonds between
          different polypeptide segments. Which one is it?

● 3-11   In a nucleic acid molecule, which carbon atoms of the deoxyribose
          bear a phosphate, a hydroxyl group, and the base?

● 3-12   Which nucleic acid base is unique to DNA? to RNA?

3-13   What is the difference between a nucleoside and a nucleotide?

● 3-14   In a nucleic acid which carbon atoms are connected by a
          phosphodiester group?

● 3-15   (a) What chemical groups are at the end of a single DNA strand?
          (b) What two chemical groups are at the end of a double-stranded
          DNA molecule?

● 3-16   What term would describe a macromolecule whose monomers do not
          engage in either attractive or repulsive interactions and which
          has free rotation between the monomers?

● 3-17   What term describes the tendency of a macromolecule to fold so
          nonpolar groups cluster?

3-18   Base stacking is a result of the insolubility of the hydrophobic
          ring portion of purines and pyrimidines. How is this insolubility
          to be reconciled with the fact that the nucleic acid bases can
          form solutions as concentrated as 0.1 M?

3-19   (a) If a particular macromolecule that is very compact in 0.01 M
          NaCl expands considerably in 0.5 M NaCl, what forces are prob-
          ably important in determining the overall size and shape of the
          molecule?
          (b) If a macromolecule that is a near-random coil and that is in
          0.2 M NaCl becomes very extended and rigid in 0.01 M NaCl, what
          forces are probably involved in the extension?

3-20   A particular enzyme loses biological activity if stored in 0.01 M

NaCl. This loss is prevented if the solution also contains 0.01 M 2-mercaptoethanol, a reducing agent. What information does this give you about the enzyme?

3-21   Electrophoresis in gels is now a common technique in molecular biology. Why is it necessary that the electrophoresis be done in solutions having low salt concentration?

3-22   In combined two-dimensional chromatography and electrophoresis, as employed in the fingerprinting technique, does it matter which is done first? Explain.

3-23   Amino acids can be separated by electrophoresis on paper. A pH is chosen to make some amino acids positively charged and others negatively charged. Thus, some amino acids will move toward the anode and others toward the cathode, their charge principally determining the direction and rate. Often, amino acids having the same charge (such as alanine and valine) also separate; in other words, they move at different rates, though in the same direction. What determines the difference in rate and will alanine or valine move faster?

3-24   Define the terms shadowing, staining, and negative contrast, as used in the electron microscopy of macromolecules.

3-25   SDS-PAGE electrophoresis of a particular protein yields a molecular weight of 52,500 with an error of only a few percent. A crude measurement of the molecular weight by a different technique known to have an error of as much as 15 percent yields the value 198,000. What is the molecular weight of the protein, assuming that the value determined by the SDS-PAGE technique is accurate?

3-26   A mixture of different proteins is subjected to electrophoresis in three polyacrylamide gels, each having a different pH value. In each gel five bands are seen.
(a) Can one reasonably conclude that there are only five proteins in the mixture? Explain.
(b) Would the conclusion be different if a mixture of DNA fragments was being studied?

3-27   An enzyme that has been extensively purified by a variety of criteria is thought to be pure—that is, it shows a single peak after chromatography in several solvents, after electrophoresis at several pH values, and after centrifugation in solutions of different ionic strength and composition. When subjected to sodium dodecyl sulfate-gel (SDS-gel) electrophoresis, two bands result, one twice the area of the other. What information does this give about the protein? Because purity is always difficult to prove, how could you prove that your hypothesis is correct? Hint: Use gel chromatography.

3-28   A virus contains 256 proteins, 64 having a molecular weight of 1800 and 192 having a molecular weight of 26,000. If the virus is disrupted and analyzed by SDS-gel electrophoresis, what would be the relative distances migrated and the relative areas of the bands?

3-29   Suppose that you have isolated a protein that seems to have two enzymatic activities. This observation makes you suspect that you may have two proteins that purify together. To check this, you subject the preparation to electrophoresis in a polyacrylamide gel at a variety of pH values. In each gel, a single band results, but the band is sufficiently broad that it may contain two proteins which do not separate. In an SDS gel, a single, broad band is also found. Because a protein with two enzymatic activities is rare, it is necessary to try a little harder to see if the breadth of the band is due to the presence of two components. What parameters could you vary to improve resolution by electrophoresis?   What other methods (nonelectrophoretic) might you try?

3-30   In the Kleinschmidt technique for visualizing DNA molecules, explain why it is necessary to use a very low angle when shadowing the DNA with metal. Is the metal being deposited directly onto the DNA molecule? Would it suffice to deposit metal from above?

3-31   Explain the principle of negative contrast in electron microscopy. A reagent commonly used in this procedure is phosphotungstic acid. In this molecule what is actually absorbing the electrons? How would the appearance of an empty phage head compare to that of a head filled with DNA?

3-32   (a) Explain how shadowing a phage sample allows the dimensions of the particle to be measured.
       (b) A spherical virus is mixed with polystyrene spheres, 705

angstroms in diameter. After shadowing with tungsten, the length of the shadow of the polystyrene is 1250 angstroms and that of the virus 820 angstroms. What is the diameter of the virus? Some of the viruses have shadows ranging from 150 to 200 angstroms and look somewhat fuzzy. What are these?

3-33   Support films for electron microscopy are always very thin layers of either plastic or pure carbon. These films are very fragile and frequently break. Stronger films could be made of metals such as chromium. Why would such metal films not be useful?

3-34   In a DNA fraction isolated from phage-infected cells, it is suspected that linked circular molecules (catenanes) may be present. In electron micrographs of this DNA, 40 percent of the observed circles appear to overlap as linked circles might. What criterion should be used to ensure that the circles are linked and not just unit-sized circles resting one on top of another on the support film? How can one minimize this kind of accidental overlap?

• 3-35   Which would have a higher sedimentation coefficient, a rigid rod or a flexible rod, both having the same radius and same mass?

3-36   An enzyme has a sedimentation coefficient of 16S at pH 7. At pH 3 the sedimentation coefficient drops to 11S and at the same time the enzyme loses all activity (even when the pH is restored to 7). It is known that there is no change in molecular weight at pH 3.
(a) What is the approximate shape of the protein at pH 7?
(b) If the first value was 4S instead and the value at pH 3 was 10S, what would you conclude?

3-37   The density of DNA in CsCl is approximately 1.7 g/cm^3 and that of most proteins is approximately 1.3 g/cm^3. What would you expect to be the density of a typical bacteriophage which is 50 percent protein and 50 percent DNA?

3-38   If the density of a typical protein is 1.300 g/cm^3, what would the density be if the protein contained ^{15}N instead of ^{14}N? ^{13}C instead of ^{12}C?

4

Nucleic Acids

●4-1 What is the base sequence of the DNA strand complementary to each of the base sequences that follow?
(a) T G A T C A G G T C G A C
(b) A T A T A T A T A T A T A T

4-2 Write down the names of the bases, ribonucleosides, deoxynucleosides, ribonucleotides, and deoxyribonucleotides in both DNA and RNA.

●4-3 Which base pair, adenine–thymine or guanine–cytosine, contains a greater number of hydrogen bonds?

●4-4 For double–stranded DNA
(a) What is the relation between the value of ([A] + [G]) and ([C] + [T]), in which the brackets indicate molar concentration?
(b) Of ([A] + [T]) and ([G] + ⌊C⌋)?

4-5 (a) What is the length-to-width ratio of a DNA molecule whose length is 20 μm?
(b) How many base pairs does the above DNA molecule have?

4-6   (a) What is the length-to-width ratio of a DNA molecule whose molecular weight is 40×10^6?

4-7   What would be the approximate molecular weight of a DNA molecule whose length is 16.4 μm?

4-8   A typical gene utilizes a segment of DNA whose molecular weight is one million. How many turns of the helix does this represent?

4-9   Indicate the 3' and 5' ends of the tetranucleotide AGAC and also write the sequence using the p notation.

4-10   Assuming that the average molecular weight of a bacterial gene is one million and that all of the DNA is used to form genes (not really the case), how many genes would there be in (a) a typical bacterium and (b) in a human cell (haploid number = 23)?

4-11   What is the value of A_{260} of a DNA solution whose concentration is 32 μg/ml?

●4-12   Order the DNA molecules shown below from lowest to highest melting temperature:

(a)   A A G T T C T C T G A A
        T T C A A G A G A C T T

(b)   A G T C G T C A A T G C A G
        T C A G C A G T T A C G T C

(c)   G G A C C T C T C A G G
        C C T G G A G A G T C C

4-13   Which of the following DNA molecules would have the lower temperature for strand separation?   Why?

(a)   A G T T G C G A C C A T G A T C T G
        T C A A C G C T G G T A C T A G A C

(b) A T T G G C C C C G A A T A T C T G
     T A A C C G G G G C T T A T A G A C

4-14 The DNA molecules shown below are denatured and then renatured. In both cases, renaturation takes place at a single temperature. Which molecule will renature with greater probability to form the original structure? Why? Hint: Think about intrastrand interactions.

(a) A T A T A T A T A T
     T A T A T A T A T A

(b) T A G C C G A T G C
     A T C G G C T A C G

4-15 The following DNA molecules are denatured and then renatured. It is found that one requires a higher temperature for renaturation. Which one is it?

(a) G A G C T G C A T C A G A T G C A G
     C T C G A C G T A G T C T A C G T C

(b) A T C G G G G T A C C C C G A T A A
     T A G C C C C A T G G G G C T A T T

4-16 A double-stranded hexanucleotide, for example, 3 guanines in one strand and 3 cytosines in the other, has very low thermal stability compared to a polynucleotide containing 1000 guanines in one strand and 1000 cytosines in the other. Explain the difference.

4-17 Suppose you had a method for measuring unpaired bases in DNA and obtained the data shown in Table 4-17 at a particular temperature. How many pairs of bases would be unpaired in a polynucleotide having 100 base pairs? Where would you expect these unpaired bases to be? Why will they be in that position? Hint: see problem 4-16.

TABLE 4-17

Number of base pairs in the DNA	Fraction unpaired (percent)
5,000	0.28
1,000	1.40
250	5.60

4-18   Table 4-18 shows the maximum solubilities of deoxyribose and the DNA bases in two solvents. In which solvent will a higher temperature be required to separate the strands of DNA?

TABLE 4-18

Solvent	Deoxyribose	DNA bases
A	4.3 molar	0.02 molar
B	0.5 molar	0.22 molar

4-19   When a DNA molecule is boiled, the strands come apart. Which of the following are. possible explanations for the strand separation? (Do not concern yourself with whether the following statements are true---only with whether they could be explanations.)

(a) The energy of thermal vibrations is greater than the energy of the weak bonds stabilizing the double-stranded structure.
(b) The solubility of adenine and guanine is greatly increased at high temperatures and the solubility of the deoxyribose is unchanged.
(c) The solubility of adenine and guanine is greatly decreased at high temperature and the solubility of the deoxyribose is unchanged.
(d) The solubility of the bases is unchanged, but the solubility of the deoxyribose is increased.

4-20   When DNA is placed in distilled water, the two strands come apart. Explain why. Hint: Consider the effect of ions in solution on the interaction between charged groups.

4-21   Consider a protein that is very polar (that is, highly charged) but has a small, very nonpolar region. The nonpolar region binds very tightly to the bases in DNA. Which of the following results would follow addition of this protein?
(a) The DNA would precipitate.
(b) The two strands would come apart.
(c) The DNA would become more stable.

4-22   When DNA is heated to 100°C, the two strands separate. When cooled again, in general, the two strands do not come back together again, because they do not collide with one another in such a way that complementary base sequences can form. In other words, heated DNA remains as single strands. Consider a DNA molecule in which the adenines and thymines alternate in each strand (i.e., ...ATATAT...). If such a molecule is heated to 100°C and then cooled, what might be the structure of the single strands? Assume that the concentration is so low that the two strands never find one another.

4-23   If DNA having an absorbance at 260 nm (A_{260}) equal to 1.00 is boiled, the value of A_{260} increases to 1.37. This is an indication of the conversion of double-stranded to single-stranded DNA. When returned to 25°C, the value A_{260} remains the same, if the DNA is in a solution of low ionic strength and contains no polyvalent cations (for example, 0.01 M NaCl), but drops to 1.12 if the ionic strength is high (for example, 0.5 M NaCl).
(a) Explain the observation.
(b) By various chemical means (e.g., by treatment with nitrous acid) the two bases in a base pair can be covalently joined. Predict the changes in A_{260} when this DNA is boiled in 0.5 M NaCl and then returned to 25°C.
(c) Predict A_{260} changes for a double-stranded deoxynucleotide containing G in one strand and C in the other. Would the change be the same for a circular polymer? Assume the ionic strength is high.
(d) A double-stranded DNA molecule whose base sequence in one strand is ATGCATATATGCAT is boiled in 0.5 M NaCl and then returned to 25°C. What will be the ratio of A_{260} before boiling and after a cycle of boiling and cooling?

4-24   Denaturants can be ranked in terms of the number of degrees of lowering of T_m by addition of the substance to yield a

concentration of 0.1 M (even though the decrease might be only a few degrees). For each of the following pairs, which substance is probably the more effective denaturant and why?
(a) Ethanol, propanol.
(b) Urea, dimethylurea.
(c) Dimethyl ether, diethyl ether.
(d) 1-amino-2-hydroxypropane, 1,3-diamino-2-dihydroxypropane.

4-25   If DNA in 0.01 M NaCl is heated to 100°C for three minutes and then cooled, the value of A_{260} is normally 1.37 times the value before heating.
(a) If the DNA is exposed to nitrous acid prior to heating, the value of A_{260} after heating and cooling is the same as that before cooling. Explain.
(b) If the nitrous acid-treated DNA is heated to 100°C for thirty minutes and then cooled, the value of A_{260} observed after cooling is about 15 percent greater than the value before heating. Explain.

4-26   If DNA is put into 100 percent methanol, its absorbance increases by 37 percent. If the methanol concentration is reduced 20-fold by dilution of the methanolic solution with water, normal absorbance and a double-stranded structure is restored. Explain.

4-27   At pH 12 strands of DNA separate because hydrogen bonds break. What is the cause of the breaking of hydrogen bonds? Why are the hydrophobic forces unable to maintain the double-stranded structure?

4-28   It has been stated that a single-stranded polynucleotide is helical if bases are stacked and this stacking tendency has been used to explain the stability of DNA. Since stacking in a single-stranded polynucleotide is between adjacent bases, how can stacking explain the tendency of the two strands of the double helix to stay together? (The principle here is the same as that in problem 4-27).

4-29   When heating double-stranded DNA, the maximum value of A_{260} reached is usually 37 percent higher than the A_{260} at 25°C. However, in a solution of 7 M sodium perchlorate, the maximum value of A_{260} reached is only 26 percent higher than that at 25°C. Explain why this is so.

4-30   Suppose you have obtained a melting curve having the normal
       shape, but the total increase in absorbance is only 20 percent.
       How might you explain the low value? Would the determination of
       T_m from such a curve be believable?

4-31   You have just prepared two DNA samples, one native and one
       denatured, and have dialyzed each against 0.01 M NaCl. You do not
       know the A_{260} values before dialysis. You then add a very small
       amount of an enzyme to each and in so doing mix up the samples.
       How can you determine the identity of each sample by an
       absorbance measurement? You may assume that a small part of each
       sample is consumed in the testing and that the enzyme will not
       interfere with the test. Design a second test that does not
       require the introduction of an agent or condition that causes
       denaturation.

4-32   Why does salt concentration affect T_m? Hint: see problem
       4-20.

4-33   What feature, if any, of a melting curve is affected by the
       molecular weight of a DNA molecule?

4-34   When melting curves for DNA solutions are obtained, the indicated
       temperature is generally that at which the absorbance (A_{260}) was
       measured. Another kind of melting curve can be obtained by
       heating the DNA to a temperature T, cooling the DNA to 25°C, then
       measuring the value of A_{260} at 25°C and plotting this value of
       A_{260} versus T. Such a melting curve is called an irreversibility
       curve or i-curve since an increase in A_{260} occurs only if a DNA

FIGURE 4-34

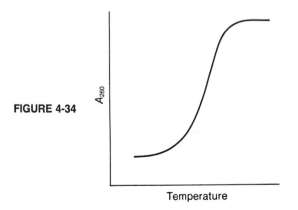

Temperature

molecule is irreversibly denatured, that is, if it remains single-stranded. For the melting curve shown, draw very carefully a hypothetical i-curve for randomly fragmented bacterial DNA, intact phage DNA, and phage DNA containing several phosphodiester breaks in the single strands.

4-35   In preparing hybrid DNA by renaturation of denatured DNA molecules of different types, some unrenatured DNA usually remains. Describe several procedures that could be used to eliminate this DNA, if it was unwanted.

4-36   When a DNA molecule is heated, base pairs are disrupted; the amount of disruption increases with temperature. At a critical temperature, the two strands separate. Draw a graph showing how the sedimentation coefficient s changes with temperature and indicating the position of T_m. Assume that the DNA molecules are in a solution of high ionic strength so that single-stranded DNA molecules are flexible. Denote the s-value of DNA at 25°C as s, and that at higher temperatures as s_T.

4-37   Bacterial DNA can be density-labeled if the bacteria are grown in a medium containing a heavy isotope (such as ^{15}N or ^{13}C). If both strands are so labeled, the DNA is said to be heavy (HH) as opposed to normally light (LL) DNA. If equal amount of HH and LL DNA are mixed, heated to 100°C, and slowly cooled to allow renaturation to occur, 25 percent of the DNA will be HH, 25 percent will be LL and 50 percent will be HL (hybrid). Suppose 45 micrograms of HH DNA and 5 micrograms of LL DNA are mixed, heated to 100°C, and renatured, how many micrograms of HH, HL, and LL will result?

4-38   DNA from organism A, labeled with ^{14}N, is hybridized with an equal concentration of DNA from organism J, labeled with ^{15}N, and then centrifuged to equilibrium in CsCl. Five percent of the total renatured DNA has a hybrid density. What fraction of the base sequences are common to the two organisms?

4-39   Double-stranded DNA is converted to single-stranded DNA at high pH (11.3 to 11.6), depending on the DNA. If the DNA contains 5-bromouracil (Bu), the critical pH is about 10.5. This conversion can be detected by centrifugation in CsCl since the DNA increases in density by 0.014 g/cm^3 when converted to single strands. Normal DNA and Bu-DNA can also be distinguished in CsCl since the density of the latter is about 0.1 g/cm³ greater than the former. At one time it was claimed that *E. coli* DNA is four-stranded and that it consists of two double helices held together by "biunial" bonds. Thus, the DNA of hybrid density observed by Meselson and Stahl was believed to consist of one LL duplex bonded to one HH duplex. Using growth for one generation in medium containing 5-bromouracil in order to obtain hybrid DNA, Robert Baldwin and Eric Shooter were able to distinguish BuL from BuBu:LL DNA (that is, four-stranded DNA containing BuBu and LL). Draw the curves for density versus pH expected for BuL and BuBu:LL DNA. (See problem 4-37 for definition of H and L).

4-40   The technique of nearest-neighbor analysis allows determination of the frequency with which a given nucleotide is adjacent to each of the four nucleotides in enzymatically synthesized DNA. Synthesis is carried out with three unlabeled deoxynucleoside triphosphates and the fourth labeled with ^{32}P in the α position. The labeled synthetic DNA is then degraded using nucleases that cleave on the 5' side of each phosphoryl group, thereby yielding 3' mononucleotides. Thus, an α-^{32}P that came in with dATP, for example, is left attached to the nucleotide that was adjacent to this A residue on the 5' side, as follows (arrows indicate points of nuclease cleavage):

$$pXPApYpZ \longrightarrow XP + Ap + Yp + ...$$

Therefore, the relative amounts of radioactive label recovered as Ap, Gp, Cp, and Tp indicate the relative frequencies of the dinucleotides ApA, GpA, CpA, and TpA in the synthesized DNA, respectively. Using another labeled triphosphate in the synthetic reaction, for example, α-^{32}P-dTTP, the frequencies of ApT, GpT, CpT, and TpT can be obtained, and so on, for each of the 16 possible dinucleotides. For a particular DNA molecule, some of the nearest-neighbor frequencies are ApG, 0.15; GpT, 0.03; GpA, 0.08; TpT, 0.10. In each case the nearest neighbor is written in the 5' → 3' direction.
(a) What would be the nearest-neighbor frequencies of CpT, ApC, TpC, and ApA?
(b) If the strands of DNA were parallel instead of antiparallel, which nearest-neighbor frequencies would you know and what would they be?

4-41   A linear phage DNA molecule is hybridized with a molecule that is identical except that the central 10 percent of the molecule is deleted.

(a) What is the structure of the heteroduplex that would be seen by electron microscopy?

(b) If the missing 10 percent is replaced by a piece of non-homologous bacterial DNA having the same length, what will the structure of the heteroduplex be? How will this differ if the non-homologous bacterial DNA is only half the length of the deleted phage DNA?

4-42   When two nearly identical samples of DNA (such as DNA molecules from mutant and wild-type virus X) are mixed, denatured, and reannealed, homoduplexes and heteroduplexes are formed as shown in Figure 4-42A. Homoduplexes contain two strands from the same sample of DNA and heteroduplexes contain one strand from each of two different samples of DNA. Sequence differences between the two DNA samples lead to noncomplementary regions in heteroduplexes that cannot hydrogen-bond and therefore remain single-stranded. If these regions are longer than 50 to 100 nucleotides, they appear as loops when the DNA is examined by electron microscopy. Two common heteroduplex structures are shown in Figure 4-42B. Examination of heteroduplex DNA is a powerful method for mapping the locations of large deletions, additions, and substitutions. From the data in Figure 4-42C construct a map of the DNA of the wild-type organisms, showing positions of segments deleted from the mutants. (Figures on pages 36-37.)

4-43   Many DNA molecules have long tracts of pyrimidines in one of the strands (and of course, purine tracts in the other). If the DNA is denatured in the presence of a polyribopurine and then centrifuged to equilibrium in a buoyant solution of cesium sulfate, the denatured DNA frequently forms two bands. It is known that the density of RNA is approximately 0.100 g/ml greater than DNA. Explain why there are two bands, what material is in each band, and whether the separation of the two bands would increase or decrease as the molecular weight of the poly-ribopurine increases.

4-44   A DNA molecule is prepared in which one strand contains ^{15}N and ^{13}C atoms ("heavy") and the other contains ^{14}N and ^{12}C atoms ("light"). This DNA is double helical, circular, and closed. Show how this material will be distributed after CsCl equilibrium density centrifugation at neutral pH if the DNA is treated

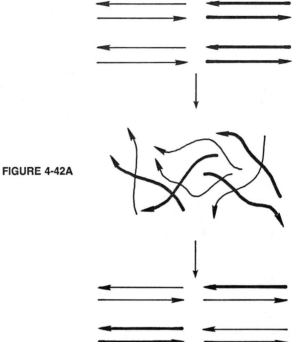

FIGURE 4-42A

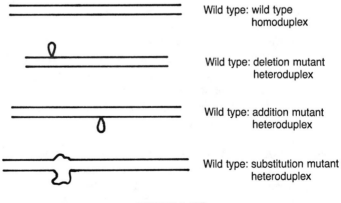

Wild type: wild type homoduplex

Wild type: deletion mutant heteroduplex

Wild type: addition mutant heteroduplex

Wild type: substitution mutant heteroduplex

FIGURE 4-42B

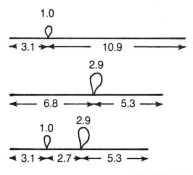

FIGURE 4-42C

in the following ways: (a) Exposure to alkali (pH 13) or heating in boiling water. (b) Treating with sufficient endonuclease to introduce two or three breaks in each strand. (c) As in (b), but followed by exposure to alkali or boiling.

4-45 Consider a long linear DNA molecule, one end of which is rotated four times with respect to the other end, in the unwinding direction. The two ends are then joined. If the molecule is to remain in the underwound state, how many base pairs will be broken? If the molecule is allowed to form a supercoil, how many nodes will be present?

4-46 Suppose ten protein molecules are bound to a circular DNA molecule having a nick. Each bound protein molecule breaks one base pair. The nick is then sealed with DNA ligase (an enzyme that seals 3'-OH—5'-P nicks) and the protein molecules are removed by treatment with a protease. What will be the shape of the DNA molecule after removal of the protein?

4-47 A covalently closed circular DNA molecule that is relaxed (neither positively nor negatively supercoiled) is partially denatured and 50 base pairs are broken. How does the degree of supercoiling change?

●4-48 Describe several ways in which a linear DNA molecule can form a circle.

●4-49 State which of the following operations will induce formation of a supercoil: joining the ends of a linear DNA and doing nothing else; twisting the two ends of a linear DNA and then joining the ends together; joining the ends of a linear DNA and then twisting the circle.

4-50 DNA molecules are easily broken by hydrodynamic shear forces. These forces act in such a way that a DNA molecule breaks near its middle. By slowly increasing the magnitude of the force, a molecule can be successively broken into halves, then quarters, then eighths, and so on. Let us call the values of a force just sufficient to break a linear molecule of mass m into halves $F_{1/2}$; for quarters and eighths we use the higher values $F_{1/4}$, $F_{1/8}$; and so on. If you have a circular molecule of mass m, approximately what force is needed to break the molecule? What is the size of the fragments?

4-51 A Hershey circle is formed by heating a linear DNA molecule that has single-stranded cohesive ends to a temperature that favors joining of the ends. In DNA solutions at very low concentrations, only Hershey circles are formed. What other structures might be expected in solutions with high DNA concentrations?

4-52 Why does a supercoil have a higher density in buoyant CsCl than a linear molecule, when ethidium bromide is present?

4-53 Would you expect the density decrease in CsCl induced by the presence of ethidium bromide to be affected by the G+C content?

4-54 In CsCl containing ethidium bromide, a supercoiled and a linear DNA have densities 1.592 and 1.556 g/cm³, respectively. What would be the density of a supercoil linked (as in a chain) with an open circle of (a) identical mass or (b) twice the mass? (Linked circles do exist and are called catenanes).

4-55 Endonuclease S1 makes a strand break only with single-stranded and not with double-stranded DNA. However, S1 can cleave supercoiled DNA, usually making only a single break. Why does this occur?

4-56 Might there be a density shift due to binding of a non-intercalating substance to a phosphate group or the deoxy-ribose? Under what circumstances can the nonintercalating substance be used to separate supercoiled from linear DNA? (Thinking about problem 4-55 will help to answer this problem.)

4-57 A sample of DNA gives one sedimentation band in CsCl but two in CsCl containing ethidium bromide. The ratio of the areas of the denser to the lighter band is 2:1. The molecular weight of the DNA is 30 million. If the DNA is treated with an enzyme that produces, on the average, one single-strand break in each molecule, what would be the ratio of the band areas after such a treatment? (Remember to use the Poisson distribution to determine the fraction of molecules that are unbroken).

4-58 The sedimentation velocity properties of a supercoiled DNA molecule are being studied as a function of the concentration of added ethidium bromide. It is found that s decreases, reaches a minimum, and then increases. Explain. Hint: s is a function of both molecular weight and shape.

4-59 Repressors are often analyzed by mixing protein solutions with radioactive DNA molecules that possess an appropriate operator, and passing the mixture through a nitrocellulose filter. The amount of bound radioactivity is a measure of the amount of repressor.
(a) What is the physical principle underlying this technique?
(b) What control must be done in order to obtain accurate values for the amount of repressor?

4-60 You are characterizing three DNA-binding proteins, A, B, and C, by studying the effect of the proteins on the sedimentation coefficient s of superhelical DNA; the results shown in Figure 4-60 (page 40) are obtained. What can you say about the binding properties of proteins A, B, and C?

4-61 In general, the viscosity η of a solution of a macromolecule increases with both molecular weight and the ratio of length to width of the molecule. The viscosity of a DNA solution decreases with time if DNase is added, owing to the production of double-strand breaks in the DNA. Single-strand breaks have no effect on viscosity. Assume that single-strand breaks are made linearly with time during exposure to an enzyme and draw a graph of the expected change of viscosity with time for a double-stranded DNA molecule which is (a) linear, or (b) circular.

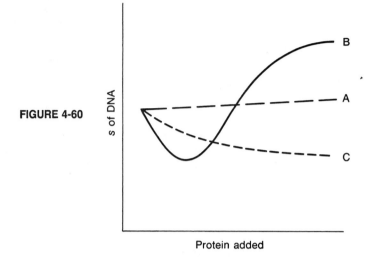

FIGURE 4-60

4-62   Double-stranded DNA can be converted to single-stranded DNA either by heating in the presence of formaldehyde or by adjusting the pH to 12.3. The sedimentation patterns of DNA denatured by either procedure are normally identical. However, if DNA is mixed with acridine orange and then irradiated with low doses of light at wavelengths absorbed by acridine orange, then the two procedures give different sedimentation patterns. With the heat-formaldehyde treatment, the sedimentation pattern of a DNA molecule is the same as that of unirradiated DNA—a single, sharp boundary with the sedimentation coefficient roughly 30 percent greater than that of native DNA; by the alkaline treatment, there is still a small, sharp boundary, but a great deal of material sediments, more slowly than this boundary. Explain this difference.

4-63   If you want to hydrolyze a double-stranded DNA molecule totally to mononucleotides, which enzyme or enzymes would you choose?

4-64   Why do you think most nucleases fail to work in 1 M NaCl?

4-65   Pancreatic DNase is a potent nuclease active on both double-stranded and single-stranded DNA, yet it rarely hydrolyzes DNA to mononucleotides. The fraction of resistant DNA decreases as enzyme concentration increases or with decreasing values of the initial DNA concentration (even if the ratio of enzyme to DNA is unchanged). If the enzymatic hydrolysis is carried out in dialysis tubing (semipermeable tubing through which small molecules but not DNA can pass) that is suspended in a huge buffer reservoir, the hydrolysis is complete. Explain.

4-66 You are studying a DNA molecule which by many standard criteria is single-stranded: it reacts rapidly with formaldehyde, it shows no increase in absorbance when heated and its sedimentation coefficient is markedly lower in 0.01 M NaCl than in 1 M NaCl. However, when treated with a 5' P-specific exonuclease known to be active against single-stranded but not double-stranded DNA, only one-third of the DNA is hydrolyzed to mononucleotides. What possibilities exist for the structure of this DNA molecule? An additional fact learned later is that if the DNA is boiled in 0.01 M NaCl and then cooled, it can be totally hydrolyzed by the exonuclease, whereas if it is boiled in 1 M NaCl, again only one-third is hydrolyzed. What can you say now about its structure?

4-67 The following two polyribonucleotides are treated with various enzymes as indicated.
(a) ApCpGpCpUpUpCp: snake venom phosphodiesterase, spleen phosphodiesterase, ribonuclease T1.
(b) ApUpC: snake venom phosphodiesterase, spleen phosphodiesterase, pancreatic ribonuclease.

Predict the digestion products.

4-68 The enzyme S1 nuclease acts preferentially on single-stranded DNA.
(a) The activity against double-stranded DNA molecules is weak but increases with temperature even when the temperature is much less than that producing an increase in absorbance. Explain.
(b) If a double-stranded DNA molecule has a missing nucleotide (a gap), S1 nuclease can break the phosphodiester bond opposite the gap. Explain. How will the rate of breakage be affected by increasing temperature?
(c) S1 nuclease is active against a supercoiled DNA molecule. Why? Will it make a double-strand break or a single-strand break?
(d) If a double-stranded DNA molecule is denatured and renatured, it remains a relatively poor substrate for S1 nuclease. However, if the DNA molecules isolated from a phage A are mixed with DNA molecules from a phage B that contains a point mutation (that is, one base pair in B differs from the base pair in the

corresponding position in A) and the mixture is denatured and renatured, approximately half of the renatured DNA molecules are broken by S1 nuclease. Why? Is a particular fraction of the renatured DNA subject to S1 activity? Which one?

4-69   An exonuclease is added to a solution of single-stranded DNA. Only a small fraction of the nucleotides is solubilized. Give several possible explanations for the resistance to the enzyme.

4-70   A DNA sample is treated with a nuclease and the absorbance at 260 nm is found to increase. When the absorbance has increased by 10 percent, the reaction is stopped and the DNA is sedimented. The sedimentation coefficient of the DNA is only slightly different from the untreated DNA and oligonucleotide fragments are not seen. What is a possible mode of action of this enzyme?

4-71   DNA extracted from λ phage is linear and double-stranded, but carries 5'-P-terminated, single-stranded, complementary ends (cohesive or "sticky" ends) that are 12 nucleotide residues long. The λ DNA is: (1) treated with the enzyme bacterial alkaline phosphatase; (2) treated with γ-^{32}P-ATP (adenosine triphosphate) and polynucleotide kinase; (3) allowed to stand for a week at 5°C in 0.1 M NaCl at low DNA concentration in order to allow the formation of circles; (4) treated with DNA ligase; and finally, (5) digested with spleen phosphodiesterase and micrococcal nuclease. The products of digestion are then chromatographically analyzed, and the only radioactive product is ^{32}P-guanosine 3'-monophosphate. What information can be derived from these findings?

4-72   Human adenovirus causes upper respiratory infections. This virus contains double-stranded DNA whose molecular weight is 20 million. When light and heavy strands of the DNA are separated and observed with the electron microscope, the structure shown in Figure 4-72 is seen. If the DNA single strands are pretreated with exonuclease III, then these circular forms are not seen. What can be said about adenovirus DNA structure?

4-73   Many phage DNA molecules are known to have cohesive ends in the sense that the 5' termini are extended to form two single-stranded segments of 10 to 20 base pairs, whose base

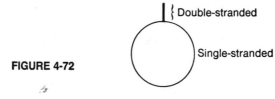

FIGURE 4-72

Adenovirus DNA structure

sequences are complementary (molecule I). Cohesive ends are easily detected by the ability of the double-stranded molecule to circularize. Molecule II would also circularize. Propose one physical and one enzymatic experiment that could distinguish the two possible structures.

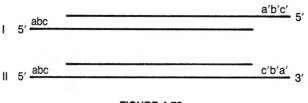

FIGURE 4-73

4-74 Explain why RNA is hydrolyzed by alkali, whereas DNA is not.

4-75 Suggest several methods by which single-stranded DNA can be distinguished from single-stranded RNA of the same molecular weight.

4-76 The usual first step in determining the base sequence of a small purified RNA molecule is to treat the RNA with T1 ribonuclease for a limited time in order to reduce the RNA to two, three, or four fragments that can be separated on an ion exchange column.
(a) How can the 5'-terminal fragment be identified?
(b) Following alkaline hydrolysis the 3'-terminal fragment yields 7 Cp, 3 Ap, 1 methyl-Ap, 3 Up, 4 Gp, 1 A (no p). What information do these data give?
(c) Treatment with venom phosphodiesterase yields 1 methyl-A (no p), 4 pA, 7 pC, 3 pU, and 4 pG. What information do these data give?
(d) Exhaustive hydrolysis with T1 RNase yields 2 Gp and three oligonucleotide fragments. Alkaline hydrolysis of each of these

fragments separately yields the following nucleotides: fragment 1—2 Ap, 1 A (no p), 4 Cp; fragment 2—1 Ap, 1 Gp, 1 Cp, 1 Up; fragment 3—1 Gp, 2 Up, 2 Cp, and 1 methyl-Ap. Which fragment is 3'-terminal?

(e) Which fragment is 5'-terminal?

(f) Hydrolysis of the T1 fragments with alkaline phosphatase and venom diesterase (not alkali, as in part (d)) yields the following products: fragment 1—1A (no p), 4 pC, 1 pA; fragment 2—1 C (no p), 1 pG, 1 pA, 1 pU; fragment 3—1 methyl-A (no p), 1 pG, 2 pU, 2 pC. What are the 5'- and 3'-terminal nucleotides in each fragment?

(g) Treatment of T1 fragment 2 with pancreatic RNase yields ApGp, Up, and Cp. What is the sequence of this fragment?

(h) How would you complete the sequence studies on the 3'-terminal nucleotide, which is nineteen bases long?

4-77   (a) A pure RNA species is degraded completely with T1 RNase, which attacks G residues and yields G-3'-P and oligonucleotides ending in G-3'-P. The degradation products are separated by two-dimensional chromatography and electrophoresis. After alkaline hydrolysis one oligonucleotide spot yields equimolar amounts of Ap, Cp, Gp, and Up. What can be deduced from this information?

(b) This oligonucleotide is treated with alkaline phosphatase to remove 5'- or 3'-terminal phosphates and then degraded completely with venom diesterase to yield pA, pG, pU, and C. What can you deduce from this?

(c) When the original T1-RNase fragment is treated with pancreatic RNase, Cp and Gp are released plus a dinucleotide that, after alkaline hydrolysis, releases Ap and Up. What is the complete sequence of the original T1 RNase fragment? Note: Venom diesterase begins at the 3'-OH ends and releases 5'P mononucleotides. Pancreatic RNase attacks pyrimidine residues, releasing C-3'-P, U-3'-P, and oligonucleotides ending in Cp or Up.

4-78   A purified RNA molecule is digested completely with T1 ribonuclease, and the oligonucleotide products are separated. Alkaline hydrolysis of one such fragment yields 1 adenosine, 1 Up, and 2 Cp residues. Treatment of the same oligonucleotide with polynucleotide kinase and γ-^{32}P-ATP, followed by complete cleavage with venom phosphodiesterase yields radioactive pU and nonradioactive pC and pA.

(a) Where in the tRNA molecule is the oligonucleotide located?

(b) What is the sequence of the oligonucleotide?

4-79   The nucleotide sequence of a single-stranded DNA is determined by the Maxam-Gilbert procedure. A sample of the DNA is given one of four different treatments.

1. The DNA is reacted with dimethyl sulfate, which methylates both G and A, though the reaction with G is fivefold faster. The DNA is then heated; the heating removes the methylated G and A residues. Then alkali is added and the sugar to which the methylated base was attached is cleaved from both neighboring phosphate groups.

2. After dimethyl sulfate treatment as in step 1 the DNA is placed in dilute acid; this preferentially removes A. The DNA is then exposed to alkali to cleave the phosphates.

3. The DNA is reacted with hydrazine and then with piperidine. This results in sugar-phosphate breakage as in treatments 1 and 2 but instead with equal frequency at sites of T and C.

4. Treatment 3 is carried out in the presence of 2 M NaCl; this suppresses the reaction with T.

These reactions are carried out only to the extent that one base in fifty is altered by the methylation or hydrazine reactions. The hydrolytic steps are carried to completion. Each reaction mixture consists of a series of fragments whose sizes are determined by the location of particular bases. Electrophoresis of each mixture through a polyacrylamide gel separates the fragments by size. The DNA is usually terminally labeled with ^{32}P in order that the fragments in the gel can be located by autoradiography. The sequence is read from the gel by observing the intensity of the bands at each position. Note that the position tells where the base is, but the intensity is a function of the probability of cleavage. Data for a particular sample are shown in Figure 4-79 (page 46).

(a) Why is the initial reaction designed to terminate when one in fifty bases have reacted; in other words, why is one in two or one in one not chosen?

(b) What is the sequence of the two DNA samples shown in panels i and ii ?

(c) A DNA molecule much longer than those in (b) can be sequenced. However, often all of the bands cannot be resolved on a single gel. To get around this, the reacted samples are divided into two parts, and each is treated by electrophoresis but for different lengths of time. What is the sequence of the DNA shown in panel iii?

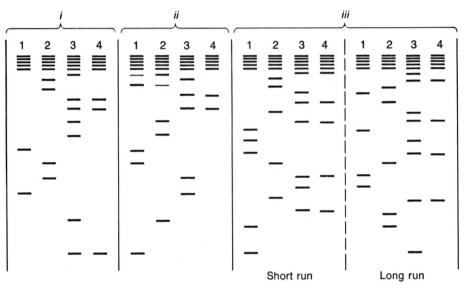

FIGURE 4-79

4-80   A DNA fragment is isolated by cutting a large DNA molecule with the HindIII restriction enzyme, which makes cuts in a particular base sequence at positions show by the arrows in Figure 4-80A. The sequence of each strand is determined by labeling the 5' ends with ^{32}P (using the polynucleotide kinase reaction) and then using the Maxam–Gilbert sequencing method. The mixture of cleaved fragments is electrophoresed yielding the data of Figure 4-80B. (The migration of the fragments is downward in the figure and the smaller fragments move the greater distance.) Observe the spacings in the ladder very carefully and answer the following questions.

(a) What is the sequence of strand A?

(b) What is the sequence of strand B?

(c) What part of the sequences could have been predicted from the information given in Figure 4-80A?

(d) Line up the complementary parts of the two strands as best you can. What irregularity can you see?

(e) From what you know about unusual bases occasionally observed in DNA, suggest an explanation for the irregularity in part (d).

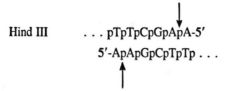

Hind III   . . . pTpTpCpGpApA-5′

5′-ApApGpCpTpTp . . .

FIGURE 4-80A

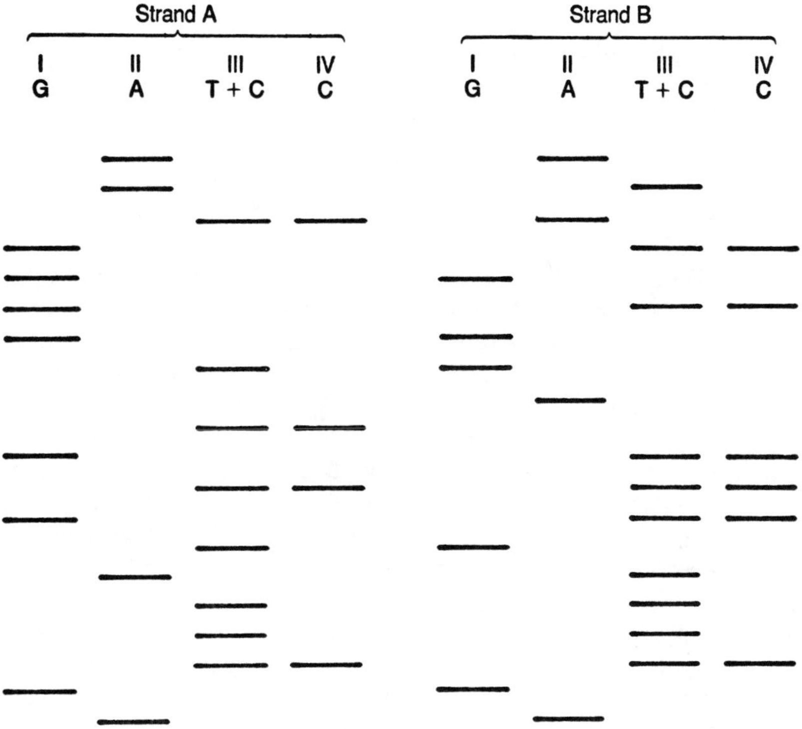

Strand A					Strand B			
I	II	III	IV		I	II	III	IV
G	A	T + C	C		G	A	T + C	C

FIGURE 4-80B

4-81   You are given a homogeneous nucleic acid sample with the
        following properties: (1) It reacts with formaldehyde; (2) upon
        denaturation its buoyant density in CsCl increases by 0.010
        g/cm³; (3) the DNA is not susceptible to either *E. coli*
        exonuclease I or spleen phosphodiesterase, even after alkaline
        phosphatase treatment; (4) its sedimentation coefficient is
        somewhat higher in high ionic strength than low ionic strength;
        (5) it binds to a nitrocellulose filter; (6) at low ionic
        strength, the sedimentation coefficient is too high for its

molecular weight, if it were simply double-stranded DNA; (7) centrifugation of the denatured material gives rise to two bands, A and B. Exposure of purified B for a short period of time to pancreatic DNase results in a molecule C with a single break. The sedimentation coefficient of C is about the same as B at high ionic strength, but lower than B at low ionic strength. The sedimentation coefficient of C is higher than A at low and high ionic strength. What is the structure of the nucleic acid?

4-82   A molecule A of a phage nucleic acid has the following properties: (1) A does not react with formaldehyde and it passes through a nitrocellulose membrane filter; (2) during centrifugation under alkaline conditions three sedimenting species are observed; (3) exposure of A to *E. coli* exonuclease III gives rise to species B; (4) exposure of B to DNA ligase results in conversion to C; (5) C sediments faster than A under alkaline conditions and also has a lower buoyant density than A in CsCl containing ethidium bromide; (6) denaturation of A, followed by its interaction with poly(G), gives three bands in CsCl; (7) exposure of each molecule of A to polynucleotide kinase in the presence of γ-^{32}P-labeled ATP results in the radioactive labeling of three nucleotides; (8) the labeling of the nucleotides in (7) takes place only if A is pretreated with *E. coli* alkaline phosphatase; (9) exposure of A to ethidium bromide results in an increased viscosity and a decreased sedimentation coefficient. What are the structures of A, B, and C?

4-83   A phage mutant has been isolated whose DNA has the following properties: (1) DNA isolated from the phage sediments as a single component (structure A) with a sedimentation coefficient that is relatively independent of the ionic strength; (2) molecule A does not react with formaldehyde, and acridine compounds cause the viscosity to increase and the sedimentation coefficient to decrease; (3) sedimentation at pH 12.5 reveals two sedimenting components and twice as much material (measured by absorbance) sediments in the faster component than the slower component; (4) when A is heated to 10°C below the melting temperature and cooled rapidly, two components B and C are produced, which have different sedimentation coefficients—when cooled slowly, only one sedimenting component is seen and it has the same sedimentation coefficient as A; (5) limited exposure of A to exonuclease III, followed by annealing, results in a structure D that has a higher sedimentation coefficient than A; (6) treatment of D with DNA ligase produces E, which in alkali sediments as a single component at least four times faster than the larger, denatured component of A; (7) when poly(G) is added to heat-denatured A, two bands differing in buoyant density are seen in CsCl density gradients; (8) if A is treated with DNA ligase, which joins a 5'-P group and a 3'-OH group, then terminally labeled with ^{32}P by the polynucleotide kinase reaction, and then digested to mononucleotides, two radioactive nucleotides, dAMP and dGMP, are found. If after the labeling but before digestion the strands are fractionated as in (8) above, the labeled dGMP can be shown to be associated with the strand that binds more poly(G). What is the structure of the phage DNA?

4-84   A nucleic acid molecule having a G+C content of 50 percent has the following properties: (1) The denaturation temperature and buoyant density in CsCl are higher than that of *E. coli* DNA; (2) the molecule is not cleaved by RNase; (3) a single scission by DNase gives rise to structure A, having a lower sedimentation coefficient than the untreated molecule; (4) denatured A (which is called B) gives rise to two bands when centrifuged to equilibrium in neutral CsCl, but only one in alkaline CsCl; (5) exposure of the original molecule to pH 13 for 60 minutes at 37°C and reneutralization gives rise to a structure C that has a lower buoyant density in neutral CsCl than before alkaline treatment; (6) molecule C is resistant to snake venom phosphodiesterase; (7) when C plus the pH 13-digestion products are filtered through a nitrocellulose membrane filter, half of the material, which is D, passes through the filter—the material retained on the filter can be eluted and is found to be C; (8) polynucleotide kinase does not phosphorylate either forms C or D. What are the structures of the parent molecule and of its derivatives?

5

Proteins

- 5-1 (a) Which amino acids are polar and which are nonpolar?
  (b) Is isoleucine more nonpolar than alanine? Why?

- 5-2 Roughly how many amino acids are contained in a protein whose molecular weight is 32,000?

- 5-3 About which bond in the polypeptide backbone is there no free rotation?

5-4 Which of the molecules—polyalanine and polyaspartic acid—will have its three-dimensional shape altered by a change in pH and what changes will occur?

5-5 Name the kinds of bonds in proteins in which the side chains of each of the following amino acids might participate: cysteine; arginine; valine; aspartic acid.

5-6 In aqueous solution, polymers containing both polar and nonpolar regions tend to fold with the nonpolar regions inside the folded structure, thus avoiding contact with water. A protein is such a polymer because some amino acids have polar and some have nonpolar side chains. Which amino acids will tend to be internal and which will tend to be on the surface of the protein? Which show little preference? Some of the polar amino acids are often found inside the protein but generally in clusters. Will cysteine also show a preference? How does cysteine differ from all other amino acids with respect to its interaction with another amino acid?

• 5-7 State whether the following statements about proteins are true or false.
(a) In general, in aqueous solution alanine is more likely to be internal than external.
(b) Serine is more likely to be internal than external.
(c) Serine is likely to be located in the active site of an enzyme because of its OH group.
(d) Polyalanine is more extended in 0.3 M NaCl than in distilled water.
(e) Polyglutamic acid is more extended in 0.3 M NaCl than in distilled water.
(f) There is free rotation adjacent to a peptide bond.
(g) All naturally occuring proteins contain at least one residue of each amino acid.
(h) A single amino acid change necessarily causes gross changes in protein structure.

5-8 What kinds of three-dimensional configurations might the following peptides have:
(a) Gly-Asp-Met-Ala-Ala-Glu-Val.
(b) Gly-Phe-Ile-Gly-Asp-Phe-Gly.
(c) Pentaphenylalanine.
(d) Decaglutamic acid.
How would you expect any of these structures to vary with pH?

5-9 What would you guess to be the environment of a glutamine that is internal? What is the environment of an internal lysine?

5-10 Most enzymes lose all activity when treated with formaldehyde. Why? Hint: Think about the chemical reactivity of formaldehyde.

5-11 The two least frequent amino acids in protein X are tryptophan and methionine. Protein X contains 0.50 mole percent tryptophan and 1.50 mole percent methionine.

(a) What is the minimum number of amino acids per molecule of protein X?

(b) If protein X has a sedimentation rate that is slightly less than that of a protein having 267 amino acids, what is the probable number of methionines per molecule of protein X?

5-12 The enzyme carboxypeptidase can remove the terminal amino acid of a protein by cleaving the peptide bond formed between the terminal amino acid and the penultimate amino acid. After the terminal amino acid is removed, the penultimate one can be removed. A protein is treated with carboyxpeptidase for a limited period of time and only alanine is recovered, in the ratio of two alanines per protein molecules. What is the amino acid sequence of the terminal tripeptide?

● 5-13 Which of the following sets of three amino acids are probably clustered within a protein? (1) Asn, Gly, Lys; (2) Met, Asp, His; (3) Phe, Val, Ile; (4) Tyr, Ser, Lys; (5) Aly, Arg, Pro.

5-14 The amino and carboxyl groups of most amino acids are hydrogen-bonded to water. In a dipeptide the peptide group is also hydrogen-bonded to water. For which of the following structures are the peptide groups not bound to water? α helix, β structure, random coil.

5-15 (a) What is the structure of polyarginine at pH 7?
(b) How might the structure be changed?

5-16 The peptide shown below is treated with trypsin.
(a) Between which pairs of amino acids does cleavage occur?
(b) How many peptides would result from a mutant in which lysine is replaced by alanine?

N–Ala–Pro–Ser–Arg–Thr–Gly–Glu–Leu–Lys–Ala–Arg–Met–Lys–COOH
        1   2   3   4   5   6   7   8   9   10   11   12   13

5-17 A protein containing twelve amino acids is cleaved by trypsin. The peptides Met-Trp-Pro-Arg, Val-Ala-Phe-Leu-Lys, and Val-Met-Gly are produced. Treatment of the original protein with carboxypeptidase yields mostly Gly and some Met. A protease that cleaves only at the amino-terminal side of Ala yields a pentapeptide and a heptapeptide. What is the amino acid sequence of the protein?

5-18 State whether the following polypeptides would aggregate to form a multisubunit protein consisting of identical subunits and give the reason for your conclusion.
(a) The folded molecule contains two distinct surface regions whose shapes are complementary and no polar amino acids are nearby.
(b) There is a large hydrophobic cluster that is in a crevice just below the surface.
(c) A large hydrophobic patch is flanked by two lysines.
(d) The surface has a region in which positively and negatively charged amino acids alternate and are in a linear array.

5-19 A protein is treated with carboxypeptidase and then precipitated from solution. The supernatant is analyzed and found to consist of 82 mole percent valine and 18 mole percent alanine. By a variety of criteria you know that at least 99.9 percent of the original sample consists of a single protein. How do you explain the production of two amino acids by carboxypeptidase?

5-20 Which of the following bonds or interactions are essential for forming an α helix and a β structure: hydrogen bonds, van der Waals forces, hydrophobic interactions, ionic bonds?

5-21 Which of the following functional classes of proteins is more likely to be fibrous rather than globular: enzymes, regulatory proteins, structural proteins?

5-22 What is the length of an α-helical polypeptide consisting of 500 amino acids?

● 5-23 Which of the following are true statements?
(a) Enzymes affect the direction of a chemical reaction.
(b) Enzymes alter the speed of a chemical reaction.
(c) Enzymes are rarely, if ever, consumed in chemical reactions.
(d) Enzymes are always proteins.
(e) A particular enzyme can catalyze reactions involving different substances.
(f) A particular enzyme frequently carries out different types of reactions involving a single chemical group.

5-24 Is an enzyme likely to have a very rigid configuration? Explain.

5-25   What is meant by induced fit?

5-26   Name several ways by which the activity of an enzyme is regulated.

5-27   Why do most enzymes lose activity if dissolved in distilled water?

5-28   Most enzymes lose activity if a dilute solution is shaken so violently that foaming occurs. This can happen even when there is no decrease in molecular weight. Why is the activity lost?

5-29   What kinds of forces might hold together a protein containing identical subunits?

5-30   A particular enzyme has a molecular weight of 60,000. When dissolved in 7 M urea, the molecular weight drops to 20,000. When the urea is removed, the molecular weight is again 60,000 and the enzymatic activity is the same as before urea treatment. A mutant enzyme shows the same changes in molecular weight but there is no enzymatic activity either before or after the urea cycle. If equal weights of the normal and mutant enzymes are mixed and treated with urea, and then the urea is removed, what can you say about the requirements for enzymatic activity if the activity is now 50 percent of the starting value? 87.5 percent? 12.5 percent?

5-31   In general, a diploid cell containing the + and - alleles encoding an enzyme E has an E^+ phenotype. You have isolated a particular mutant that yields an E^- phenotype even when the + allele is present. Suggest a possible explanation for the observed phenotype.

5-32   An enzyme has been dissociated into four identical subunits. You want to test for the enzymatic activity of the individual subunits and thus must be sure that there are no tetramers remaining in the sample.
(a) What chromatographic procedure would you choose to free the monomers from the tetramers?
(b) How would you know where the tetramer would be, if it were present?

5-33   Many enzymes are aggregates of identical subunits. Explain the biological advantages of this phenomenon.

5-34   Why do enzymes vary so much in size? Would you expect the enzymes
        involved in glucose metabolism to be relatively large or
        relatively small? What would you expect to be the size range of
        the various DNA polymerases?

5-35   No amino acid side chain is a good electron acceptor, yet many
        enzymes catalyze reactions in which the enzyme serves as an
        electron acceptor. How is this accomplished?

5-36   The activity of many enzymes is strongly dependent on pH and is
        maximal at a particular pH. Furthermore, the range of pH values
        at which there is significant activity is often only two to three
        pH units. Frequently there is no detectable conformational change
        in this pH range, so it is unlikely that a gross conformational
        change accounts for this behavior.
        (a) Explain the dependence on pH.
        (b) What might you conclude about the active site of an enzyme
        if the activity increases from pH 5 to pH 7 and then remains
        constant (that is, it has no maximum) until pH 11, at which value
        the activity abruptly drops to zero.

5-37   An enzyme is known to require a high concentration of the
        magnesium ion for activity. If the magnesium ion is removed, the
        protein is irreversibly denatured. In attempting to develop a
        purification scheme for the enzyme, you try both ion–exchange and
        gel chromatography. In both cases the enzyme loses activity.

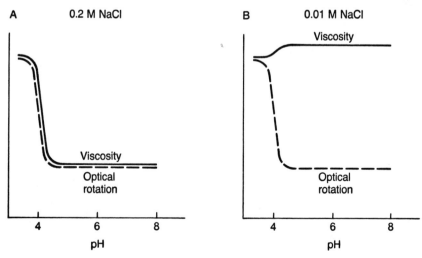

FIGURE 5-38

Explain why that might happen. In view of your explanation, what modifications could you make to improve the situation?

5-38 Polyglutamic acid was studied in (A) 0.2 M NaCl or (B) 0.01 M NaCl. Hypothetical viscosity and optical rotation data, as functions of pH, are shown in Figure 5-38 (page 55). Explain the structural alteration produced by changing the pH and the difference in the viscosity profiles. Note: optical rotation increases as helicity increases and viscosity is a measure of the degree of extension of a macromolecule.

6

Macromolecular Aggregates

6-1 Collagen fibers extracted from calves are easily dissociated by suspension in 1 M NaCl. The fibers extracted from adult cows are resistant to 1 M NaCl and cannot be dissociated except by treatment with acid. Explain the difference.

6-2 Tropocollagen is a molecule with microscropic dimensions but is capable of forming tendons many centimeters long. What feature of the molecules allows the molecule to form such long fibers?

6-3 Functional tobacco mosaic virus particles all have the same length. Virus-like particles can be reconstituted—in the absence of RNA—from purified virion protein. What relation would you expect between the lengths of the virus particles and the RNA-free reconstituted particles?

6-4 If young rats ingest seeds of the sweet pea, *Lathyrus odoratus*, they develop a condition called lathyrism in which many bodily structures lose mechanical strength; for example, the tail is easily pulled off. Lathyrism is a collagen disease. Isolated tropocollagen is normal and collagen fibers have the normal band pattern. Suggest a molecular defect.

6-5   Vitamin C is required to maintain prolyl hydroxylase, the enzyme that forms hydroxyproline, in an active form. Vitamin C deficiency causes scurvy, whose initial symptom is the appearance of bloody spots under the skin all over the body. Suggest why the red spots appear.

6-6   Why do you think that the peptidoglycan layer has evolved with so many chemically protected groups?

6-7   Why do you think bacteria evolved with a multilayered cell wall? Consider the fact that most animal cells are bounded only by a membrane similar in structure to one of the layers of the bacterial cell wall.

6-8   Suggest a function for the chromosome scaffolding protein in *E. coli.*

6-9   Define the terms chromatin, histones, nucleosome, core particle, and linker DNA.

6-10   Suggest a function for nucleosomes.

6-11   At high temperatures amines catalyze breakage of phosphodiester bonds. This phenomenon has been used to study a model system for nuclease activity by using apurinic acid as a substrate. Apurinic acid is a DNA-like molecule obtained by treatment of DNA with weak acid, which causes cleavage of the N-glycosylic bonds between purines and deoxyribose. Thus, apurinic acid lacks all purines, though the sugar-phosphate chain is intact. The enzyme analogue is a tripeptide. Some of the following tripeptides are very effective in hydrolyzing apurinic acid. Rank the following tripeptides in order of effectiveness and give reasons for your answer: Lys-Trp-Lys, Lys-Gly-Gly, Lys-Trp-Gly, Lys-Tyr-Lys, Lys-Gly-Gly.

6-12   A DNA-protein complex has been isolated and you wish to determine whether the DNA and protein are covalently linked.
(a) Which of the following experimental results would give information about covalent linkage? (1) Treatment of the complex with 2 M NaCl does not dissociate the protein from the DNA; (2) treatment of the complex with a reagent that breaks down hydrophobic interactions does not dissociate the complex; (3) digestion with a protease removes all detectable protein from the DNA.

(b) Describe an experimental result which would prove that there is covalent linkage.

6-13 A DNA-binding protein binds tightly to double-stranded DNA and very poorly to single-stranded DNA. It can bind to all base sequences with equal affinity. In 1 M NaCl binding is poor. What is the probable binding site?

6-14 A DNA-binding protein binds weakly to double-stranded DNA in 0.01 M NaCl and quite strongly in 1 M NaCl. Furthermore, it only binds to specific base sequences. What features of the protein might account for this effect of salt concentration?

6-15 If the λ Cro protein and λ DNA are mixed, the Cro-λ-DNA complex forms. What would you think would happen if an acridine compound or ethidium bromide (both of these are intercalating agents) were added to the DNA before Cro protein was added?

6-16 The folded *E. coli* genome has a high sedimentation coefficient and low viscosity. Why is this so? What treatment would both lower the sedimentation coefficient and raise the viscosity?

6-17 Many proteins consist of several identical subunits. Some of these proteins have a single binding site, whereas others have several identical binding sites. What might you predict about the locations of the binding sites in these two classes of multisubunit proteins?

6-18 When multimeric proteins consisting of identical subunits have several binding sites for the same molecule, binding is usually cooperative, there being some allosteric changes on binding. This is not the case for IgG in that each unit consisting of one L-chain and one H-chain is capable of binding antigen with equal affinity. Given the biological role for IgG, can you think of any reason for the molecule to have two identical binding sites?

7

Early Experiments in Molecular Genetics

• 7-1   State whether the following statements are true or false.
(a) In organism I a particular protein PI differs from the nearly identical protein PII made by organism II; thus, the DNA sequences that encode PI and PII must be different.
(b) One function of mRNA is to recognize codons.
(c) No two amino acids are specified by the same codon.
(d) Many single amino acids bind tightly to DNA.
(e) Many proteins bind to DNA.

7-2   An early structural model for DNA was the so-called tetranucleotide hypothesis: four nucleotides (one each of adenine, cytosine, guanine, and thymine) were thought to be covalently linked to form a planar unit. It was suggested that units are linked together to yield a repeating polymer of the tetranucleotide. Explain why, if DNA had such a structure, it would probably not have become the genetic material of cells?

7-3   Which of the following are true?
The original experiments on transformation in *Pneumococcus* showed that
(a) A hereditable character of a bacterium could be permanently

altered by exposure to DNA from a bacterium having a different character.

(b) A nonheriditable character of a bacterium could be permanently altered by exposure to DNA from a bacterium having a different character.

(c) All genetic material must contain DNA.

(d) All of the genes in a bacterium must be carried on DNA molecules.

7-4   Following publication of the transformation experiments of Avery, MacLeod, and McCarty, opponents of the DNA-gene theory argued that the transformation was caused by proteins that were contaminating the DNA sample.

(a) If transformation was indeed carried out by protein rather than DNA molecules and if the DNA preparation used contained at most 0.02 percent protein, how many protein molecules (each consisting of about 300 amino acids) would have been present in one milliliter of a DNA solution at a concentration of 10^{-7} mg/ml?

(b) If protein was the active agent in transformation, would the number calculated in part (a) account for the fact that in a typical transformation experiment 1000 transformants result from 0.0001 g of *Pneumococcus* DNA?

7-5   A criticism of the transformation experiments was that DNA might somehow be involved in the biosynthesis of the polysaccharide coat of the virulent *Pneumococcus*. Thus, the nonvirulent mutant would be lacking some step required for polysaccharide synthesis and this step could be bypassed by the addition of DNA. By this argument, DNA would not be a genetic substance. What kinds of experiments could be done to eliminate this argument?

7-6   Suppose you wish to prove to a skeptic that transformation is mediated by DNA and not by protein. You do not have available pure enzymes that degrade DNA or protein. However, you have worked hard and have shown (i) that 50 different genetic traits can be transformed, and (ii) that transformation is very inefficient and highly dependent on DNA concentration. The only piece of nonmicrobiological laboratory equipment you possess is a preparative ultracentrifuge and with it you are able both to measure the molecular weight of DNA and fractionate DNA according to molecular weight. Design a simple experiment to prove that the transforming principle is DNA. Hint: Think about linkage.

7-7   A critic of the interpretation of the transformation experiment might say that protein is the genetic substance and the protein can penetrate the cell only when the protein is bound to DNA. Thus, the loss of transforming activity following boiling of the DNA might be a result of dissociation of protein from the DNA. Assume that you have current knowledge about the denaturation of proteins and of DNA, about the effect of low and high pH on the chemical and physical properties of DNA and protein, and about the ionic strength dependence of the binding of protein to DNA; design an experiment to prove that the critic is incorrect.

7-8   Molecular biology was in a formative stage when Avery and his colleagues performed the transformation experiment. Many kinds of instruments and techniques are now available that would have simplified their investigation. Suggest a way that CsCl density gradient centrifugation might have been used to determine whether DNA or protein is the genetic material.

7-9   These questions refer to the Hershey–Chase experiment.
(a) How would you explain the fact that not all of the ^{35}S is stripped off by blendor action?
(b) Why did 30 percent of the ^{32}P remain in the supernatant after blending?
(c) What is the significance of the fact that phage are produced by the infected bacteria after blending?

7-10 Suppose that when the Hershey–Chase experiment was done, the following results were obtained. After blending and centrifugation, the ^{35}S sedimented into the pellet and the ^{32}P remained in the supernatant. Microscopic observation showed that after blending, all of the bacteria had been broken into small fragments. Electron microscopic examination of the pellet showed that there were no whole cells but just phage shells attached to small fragments of something (you do not know what). What else would you have to know, measure, or determine to enable you to draw the conclusion that DNA is the genetic material of the phage? Tell what reservations, if any, you might have about your conclusion.

7-11 (a) You wish to prove that DNA but not protein is the genetic material for a particular phage. You conceive of an experiment like the blendor experiment but unfortunately the phage has a short tail and cannot be removed from the cell wall with a blendor. You know how to isolate and purify progeny phage. You prepare phage labeled with both ^{32}P and ^{35}S, infect bacteria, and collect the progeny from the infection. Which of the following results could be taken as evidence to support your belief? Explain why.

1. ^{35}S never appears in progeny phage.

2. ^{32}P always appears in several progeny phage but the amount per phage is much less than the amount in the parental phage particles. ^{35}S appears in no phage particles.

3. ^{32}P appears in only one or two progeny phage particles and in only one strand of the DNA. ^{35}S appears in many progeny phage and the amount per phage is much less than the amount in the parental phage particle.

(b) If DNA is the genetic material, is it possible that no ^{32}P would appear in any progeny phage? Explain your answer.

7-12   When a transfer experiment is done with *E. coli* phage ϕX174, very little ^{32}P from the infecting phage appears in progeny particles. Fortunately this phage was not used by Hershey and Chase. If one insists that DNA is the genetic material, what hypotheses might be given to explain the result of this transfer experiment?

8

DNA Replication

● 8-1 State whether each of the following is true or false.
(a) In the synthesis of DNA the covalent bond which forms is between a 3'-OH and a 5'-P group.
(b) In general, the DNA replicating enzyme in *E. coli* is DNA polymerase I.
(c) A single strand of DNA can be copied if the four nucleoside triphosphates and polymerase I are provided.
(d) If polymerase I is added to the four nucleoside triphosphates without a DNA template, DNA is synthesized but with a random base sequence.
(e) An RNA primer must be complementary in base sequence to some region of the DNA to initiate DNA synthesis.

8-2 If one of the following enzymes is absent, not even one nucleotide can be added at the replication fork. Which enzyme is it? (1) Polymerase I (polymerizing activity); (2) polymerase I (5' → 3' exonuclease activity); (3) polymerase III; (4) DNA ligase.

● 8-3 From what substrates is DNA polymerized? What properties do all known DNA polymerases share?

- 8-4 What is meant by the terms primer and template?

- 8-5 What is the role of RNA in DNA replication?

- 8-6 What is the chemical group (3'-P, 5'-P, etc.) at the sites indicated by the dots labeled a, b, and c?

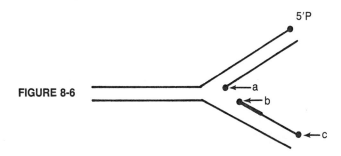

FIGURE 8-6

8-7 What is the phenotype of each of the following *E. coli* DNA mutants: *dnaA⁻*, *dnaB⁻*, *dnaC⁻*, *dnaE⁻*, *dnaF⁻*, *dnaG⁻*, *polA⁻*, *polB⁻*, and *lig⁻*, as far as DNA synthesis is concerned?

8-8 Consider the Meselson-Stahl experiment in which *E. coli* is grown for a long time in ^{15}N medium and then transferred to ^{14}N medium. (a) Assuming that the bacterial DNA is fragmented into at least 100 pieces during isolation and that replication is semi-conservative, what fraction of the total DNA is found at the density of $^{15}N^{15}N$, $^{15}N^{14}N$, and $^{14}N^{14}N$ DNA at 0, 1/2, 1, and 1-1/2 generations after transfer of the *E. coli* to ^{14}N medium? (b) What would the distribution be if the DNA is not broken but remains as a single unit? (c) Answer (b) for the case of conservative replication.

8-9 If, in a procedure such as that in problem 8-8 the data in Table 8-9 were obtained, what conclusion could you draw about the replication cycle?

8-10 Consider problem 8-8 again. At 0 generations, when the medium is changed from ^{15}N to ^{14}N, a substance X is added. After approximately 1/2 generation, 32 percent of the DNA is hybrid. After many hours, still only 32 percent of the DNA is hybrid. What does X probably do? (Note: The significance of the value of 32 percent may escape you. This is an actual value, though the main point that you should consider is that, whereas some DNA

TABLE 8-9

	Percent of total DNA		
Generation	Heavy	Hybrid	Light
0	100	0	0
$\frac{1}{2}$	34	66	0
$\frac{3}{4}$	25	75	0
$\frac{7}{8}$	6	82	12
1	0	80	20

becomes converted to hybrid, the amount never exceeds this value.) Give an example of a type of substance that might behave like X.

8-11 Consider a Meselson-Stahl density-label experiment, as described in problem 8-8, in which cells are grown for a long time in ^{15}N medium and then transferred to ^{14}N medium for one generation. Assume that you are sure that the isolated DNA is two-stranded and that it is broken into approximately 100 fragments. If only $^{15}N^{15}N$-DNA and $^{14}N^{14}N$-DNA are found, which of the following are true?

(1) DNA is not replicated by the Watson-Crick semiconservative scheme.

(2) DNA is semiconservatively replicated and the DNA is four-stranded.

(3) DNA is semiconservatively replicated and the DNA is two-stranded.

8-12 It was a striking feature of the Meselson-Stahl experiment that at all times only three discrete DNA densities were observed, corresponding to fully "heavy", hybrid, and fully "light" material. Even at fractional generation times, DNA of intermediate densities was not found. However, when the chromosome is broken into about 100 fragments, surely one of these fragments contains a replication fork and should have an intermediate density. Why is this component not seen as a discrete peak?

• 8-13 Distinguish the roles of polymerases I and III in DNA replication.

8-14   What experimental result shows that primase rather than the principal RNA polymerase makes a primer for precursor fragments?

● 8-15   What are the roles of the various exonuclease activities of the DNA polymerases in DNA replication?

8-16   Why is the activity of most nucleases and polymerases inhibited by the addition of a chelating agent such as EDTA?

8-17   DNA polymerases have the job of forming phosphodiester bonds between $3'$-OH and $5'$-P groups. However, they cannot join together these two groups since the reaction requires activation. Therefore, the polymerases use nucleoside triphosphates containing a high-energy phosphate bond. However, DNA ligases are capable of joining together a $3'$-OH and $5'$-P group. How do the ligases satisfy the requirement for activation energy?

8-18   An exciting experiment of the mid-1960's was the enzymatic conversion *in vitro* of the single-strand circular DNA molecule of phage ϕX174 to a double-stranded covalent circle by mixing ϕX174 DNA, the four deoxynucleoside triphosphates, highly purified polymerase I, and a cell extract that had been boiled and contained no protein. If this extract was dialyzed, the polymerization did not occur. What was in the extract?

8-19   In a bacterium such as *E. coli*, two classes of priming events occur in the course of DNA synthesis. In one round of replication roughly how many priming events occur that utilize (a) RNA polymerase and (b) primase?

8-20   It has been stated that helicases are needed because no DNA polymerase can unwind a double helix. However, before helicases were discovered, enzymologists were able to replicate DNA using reaction mixtures that probably contained no helicases.
(a) What property of DNA allows advance of the replication fork without a helicase?
(b) What effect would you expect a helicase to have on either the rate or fidelity of replication?

8-21   Distinguish the roles of helicases and ssb proteins in DNA replication.

8-22   It has been pointed out that an intracellular pyrophosphatase
       prevents degradation of DNA by the hydrolytic activity of DNA
       polymerases. Polymerization does occur *in vitro* even if no
       pyrophosphatase is present. Why is pyrophosphatase needed *in
       vivo* but not *in vitro*?

8-23   For certain experimental purposes a highly radioactive sample of
       particular DNA molecules is needed. This can be prepared by
       copying a template using polymerase I and very radioactive
       deoxynucleoside triphosphates, if a priming system is included in
       the reaction mixture. However, most laboratories do not have
       available the complete set of elements needed for priming.
       Suggest a much more direct method to prepare such DNA using
       polymerase I as the only protein element and without the addi-
       tion of any priming oligonucleotide.

8-24   The addition of high concentrations of uridine to a growing
       bacterial culture containing thymine causes a ribosyl-exchange
       reaction forming ribosylthymine and uracil. Bacteria that are
       Thy⁻ cannot grow in a minimal medium supplemented with thy-
       mine and uridine yet Thy⁺ bacteria grow normally in the same
       medium.
       (a) Explain the difference in growth ability of the Thy⁺ and Thy⁻
       strains.
       (b) What nutrients could you supply to a Thy⁻Ura⁻ mutant so that
       it could grow?

8-25   Using a single-stranded, circular phage DNA as a template, it is
       possible to carry out synthesis of precursor (Okazaki) fragments
       in *E. coli* extracts. The reaction produces DNA chains that are
       complementary to the circular template and carry a tract of RNA
       on one end. When synthesis is carried out with all four
       α-^{32}P-labeled deoxynucleoside triphosphates, and the product is
       degraded by a procedure that yields 3'-ribonucleotides, the
       distribution of radioactivity among these four ribonucleotides is
       that shown in the first line of Table 8-25. The next four lines
       show the results when only one of the four deoxynucleoside
       triphosphates is radioactive.
       (a) Is the RNA tract on the 3' or the 5' end of each new DNA
       strand?
       (b) What else can you conclude about the linkage of RNA to DNA?
       (c) What is the minimum number of different nucleotide sequences
       on this DNA template at which polydeoxynucleotide synthesis
       initiates?

TABLE 8-25

^{32}P-labeled substrate	3'-ribonucleotide isolated			
	Ap	Gp	Cp	Up
	^{32}P atoms per growing-point fragment			
All four nucleotides	0.99	<0.01	<0.01	<0.01
dATP	0.20	–	–	–
dGTP	0.72	–	–	–
dCTP	<0.01	–	–	–
dTTP	0.08	–	–	–

8-26 Over a wide range of cell doubling times, (produced by growing the bacteria in different media) the E. coli chromosome goes through an entire cycle of DNA replication in 40 minutes. At what average rate (expressed in nucleotides per second) do the replicative forks progress along the chromosome?

8-27 The origin of the E. coli chromosome is on the genetic map shown in Figure 8-27. What is the sequence (in time) of replication of the genes A through G? State any assumptions that you have made.

FIGURE 8-27

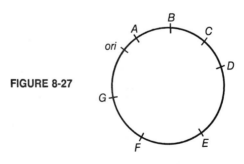

• 8-28 If a linear DNA molecule replicates with a Y structure, from which properties of DNA and the E. coli DNA polymerases does it follow that both strands cannot be replicated by growth in the same direction (as shown in Figure 8-28)? How is this problem solved by the cell?

FIGURE 8-28

8-29   How is it shown experimentally that small fragments (Okazaki fragments) are present in the region of the replication fork?

8-30   What types of information have been provided to discriminate between the following two alternatives: (1) Precursor fragments are synthesized on both sides of a replication fork. (2) One strand is synthesized as a continuous unit and the other is synthesized discontinuously as precursor fragments.

• 8-31 To join together two precursor fragments, which of the following sequences of enzymes is probably used? Assume both fragments are already made.
(1) Polymerase I (5' → 3' exonuclease), polymerase I (polymerase), ligase.
(2) Polymerase I (5' → 3' exonuclease), polymerase III, ligase.
(3) Ribonuclease, polymerase III, ligase.
(4) Primase, polymerase I, ligase.

8-32   What is the probable sequence of events used to join two uracil fragments?

8-33   The gene *dnaE* codes for the polymerase that makes DNA in *E. coli*. The only known *dnaE⁻* mutants are temperature-sensitive; that is, replication is normal at 30°C, but at 42°C, there is no DNA synthesis.
(a) Assume a culture of the cells is pulse-labeled with ³H-thymidine at 30°C. Then, the ³H-thymidine is removed from the growth medium and at the same time the cells are shifted to 42°C. What happens to the precursor fragments during the period at 42°C?
(b) Would you expect to find radioactive precursor fragments in a *dnaE⁻* mutant grown at 42°C in the presence of ³H-thymidine?

8-34   RNA polymerase can only synthesize RNA in a single direction. Suppose you have a double-stranded RNA molecule that is being replicated by RNA polymerase. Would you expect to find precursor fragments of newly synthesized RNA?

8-35 Suppose you have a DNA molecule with a gap in one strand 5000 nucleotides long and terminated with a 3'-OH group and a 5'-P group. If a DNA polymerase is added to this molecule in vitro (with everything else needed to make DNA), will the DNA filling the gap be a single piece or consist of short fragments? Would you expect fragments of any kind if the gap was filled in vivo?

8-36 In early studies of the properties of Okazaki fragments, it was shown that these fragments could hybridize to both parental strands of E. coli DNA. This was taken as evidence that they are synthesized on both branches of the replication fork. Even without any rigorous evidence contradicting the idea that precursor fragments are found in both daughter strands, a particular characteristic of the replication of E. coli DNA invalidates a conclusion based solely on the hybridization data. What is that characteristic?

8-37 Suppose you are studying a DNA polymerase similar to E. coli polymerase I. This enzyme, like polymerase I, possesses a polymerizing activity and a 5' → 3' exonuclease activity. It has other activities also but they are unknown to you. When the enzyme is in the presence of a double-stranded DNA template molecule containing a single-strand break with a free 3'-OH group and a free 5'-P group, and a mixture of the four radioactive deoxynucleoside 5'-triphosphates, radioactive DNA is detected. (This means that acid-insoluble radioactive material is found.) This is used as an indication of DNA synthesis since the nucleotides are soluble in acid and DNA is insoluble in acid. After 20 percent synthesis (that is, synthesis of an amount of acid-insoluble radioactivity equal to 20 percent of the weight of the template DNA), no increase in the molecular weight of the template DNA molecule is detectable and the density in CsCl remains that of double-stranded DNA. With a mutant enzyme, having normal polymerizing activity but lacking the exonuclease activity, after 20 percent synthesis both the molecular weight of the template DNA increases and the density of the radioactive material is greater than that of double-stranded DNA (but not as high as that of single-stranded DNA).
(a) What further property of this polymerase have you uncovered?
(b) What will happen to the density of the radioactive DNA produced by the mutant polymerase, if it is treated with an exonuclease that cleaves the molecule at the 5'-P terminus and hydrolyzes only single-stranded DNA?

(c) If the original DNA is radioactive and the triphosphates are nonradioactive, how will the amount of acid–insoluble radio-activity be affected by exposure to the wild–type enzyme and the nonradioactive triphosphates?

8-38   The central region of a linear molecule is unwound and then the ends are joined covalently, forming structure I. A portion of I is unwound and a DNA-binding protein is added to stabilize the unwound section, forming structure II. Which structure is a positive superhelix and which is a negative one?

8-39   Naturally occurring supercoiled DNA molecules have about one negative twist per 200 base pairs. When ethidium bromide is added to a sample of supercoiled DNA, binding occurs, but once the number of positive twists of the DNA reaches about one twist per 150 base pairs, no more ethidium bromide can bind.
       (a) What fraction of the molecular weight of a naturally occurr-ing supercoil can be replicated without the help of DNA gyrase?
       (b) How does this fraction compare to the size of a D loop in mitochondrial DNA?

8-40   Certain topological and energetic constraints result from the helical structure of DNA and the circularity of the DNA template.
       (a) Distinguish the roles of helix-destabilizing proteins and un-winding enzymes in DNA replication.
       (b) The *E. coli* ω protein can reduce the number of superhelical twists in naturally occurring supercoiled DNA. DNA gyrase can act on a nonsupercoiled DNA and introduce superhelical twists in the same orientation as in naturally occurring DNA. Only one of these proteins is needed for DNA replication. Which is it and why?

8-41   Which of the following are true statements? (Several are.)
       (1) There is no obvious reason for a D loop to be seen in a lin-ear DNA molecule.
       (2) If a circular DNA molecule gets a single-strand break before replication begins, it is unlikely that a D loop will be seen.
       (3) D loops must enlarge by bidirectional movement of the rep-lication forks.

8-42   The single-stranded, circular DNA molecule of phage SC3 cannot be replicated by polymerase III unless RNA polymerase and four ribonucleoside triphosphates are added along with the four deoxyribonucleotides. The product is an open circular, double-stranded DNA containing a few ribonucleotides. In order to analyze the sequence of the initiating nucleotides, deoxyribonucleoside triphosphates labeled with ^{32}P in the α-phosphate are used; the ribonucleoside triphosphates are unlabeled. The reaction product is treated with alkali and some radioactivity is released, but only into adenosine 3'-P (not the deoxy form).

(a) What sequence information can be deduced from this result?

(b) Next, α-^{32}P-dGTP is used with the other deoxynucleotides unlabeled, and the label is transferred only to adenosine 3'-P. What new sequence information is obtained?

(c) If you use α-^{32}P-dGTP as the only labeled triphosphate and no ^{32}P appears in free ribonucleotides, what sequence information can be deduced? Where does the ^{32}P label end up in this case?

● 8-43   What is the chemical group (for example, 3'-P, 5'-P, and so on) which is at the indicated terminus of the daughter strand of the extended branch of the rolling circle shown in Figure 8-43?

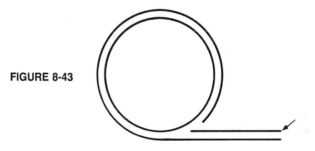

FIGURE 8-43

8-44   Answer the following questions about rolling circle replication.

(a) How many *de novo* initiation events are needed in a complete replication cycle?

(b) If the circle has unit length, what is the maximum length of the linear branch that can be formed?

(c) What is the primer?

8-45   Rolling circle replication initiates at a nick. Assume that polymerase III is the replication enzyme, discuss several events that must occur before polymerase III can initiate polymerization.

8-46   Will a donor DNA fragment, introduced into a recipient cell by transformation, generally be able to replicate in the recipient cell without integration? Explain.

8-47   Consider a phage particle containing a small linear, single-stranded DNA molecule. Its replication mode is studied by centrifugation to equilibrium in CsCl. Its density in CsCl is 1.714 g/cm^3. Phage having ^{14}C-labeled DNA infect the bacterium in medium containing ^{3}H. Samples are taken at various times, DNA is isolated, and each sample is centrifuged. The results shown in Figure 8-47 are obtained. How does this phage replicate its DNA? Do you think any progeny phage will be ^{14}C-labeled?

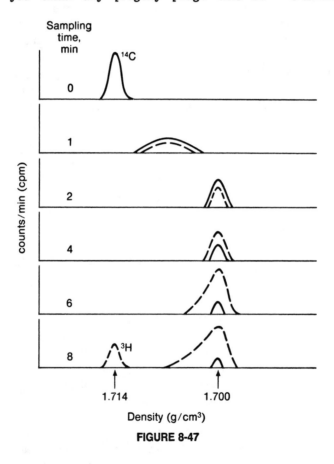

FIGURE 8-47

8-48   Initiation of a round of DNA synthesis requires RNA synthesis. *E. coli* RNA polymerase is inhibited by the antibiotic rifampicin; primase is resistant to the drug.

(a) What is the consequence, with respect to DNA replication, of adding rifampicin to a population of *E. coli* that is growing logarithmically?

(b) What is the effect of rifampicin if the cells are initially starved for a required amino acid for two hours and then both rifampicin and the required amino acid are added?

8-49   *E. coli* DNA replicates bidirectionally. The required time for transit of both replication forks is 40 minutes. Furthermore, it is known that at 37°C, even though more DNA is synthesized per unit time in a certain growth medium than in others, the rate of addition of nucleotides to the growing strand is the same in all growth media. Explain then how *E. coli* can divide every 22 minutes in some growth media.

8-50   Referring to problem 8-49, in some mammalian cells the rate of addition of nucleotides is about 5 percent of that in *E. coli*. How many growing points must there be in a mammalian cell containing three picograms of DNA per cell and replicating its DNA in six hours?

8-51   Bidirectional replication was first detected during vegetative growth of *E. coli* phage λ by electron-microsopic observation of partially denatured DNA. A partially denatured linear DNA isolated from another phage particle has the denaturation map shown in Figure 8-51. Draw a partially denatured, replicating, circular DNA molecule that would be obtained after 50 percent of the molecule is replicated, if the molecule replicates unidirectionally from a position exactly in the middle of the linear DNA. Repeat for bidirectional replication.

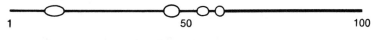

1   50   100

FIGURE 8-51

8-52   Consider the phage DNA in problem 8-51 and assume it replicates bidirectionally. In order to determine whether termination results from simple collision of the replication forks or by stopping of the forks at a genetically defined site, a phage DNA molecule is constructed in which the piece of DNA located 36 to 47 percent from the left end is duplicated in tandem sequence and the piece from 59 to 68 percent is deleted. The molecules in Figure 8-52 (page 76) were observed. Is there a defined terminus?

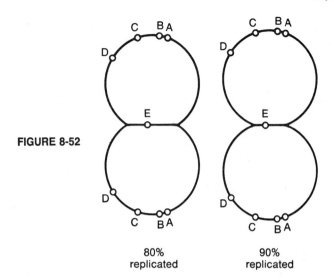

FIGURE 8-52

80%
replicated

90%
replicated

8-53   What kinds of recognition sites on DNA are required for unidirectional replication, bidirectional replication, and replication of a single-stranded circular DNA?

8-54   Bidirectional replication of eukaryotic DNA has been demonstrated by pulse–labeling with ^{3}H-labeled thymidine and observing the DNA by autoradioagraphy. The result is shown in Figure 8-54.
(a) How would the autoradiogram appear if replication were unidirectional?
(b) What would be the appearance if there were multiple unidirectionally replicating units?

FIGURE 8-54

8-55   *E. coli* strain A has a temperature-sensitive (*dna*$^-$(Ts)) mutation in a gene involved in the initiation of replication; the strain is also Lac$^-$. At 42°C no colonies can form when the bacteria are plated on agar. Strain B is wild type and contains the transmissible sex factor *F'lac*. Strains A and B are mated and cells having the phenotype Lac$^+$Dna$^-$(Ts) (strain C) are selected. These also fail to grow at 42°C, and segregate Lac$^-$Dna$^-$(Ts) cells at a frequency of about 0.8 percent per generation. At 42°C although colonies from a culture of A arise at a frequency of 10^{-6}, temperature-resistant revertants of strain C arise at a frequency of 10^{-4}; these survivors also lose the property of segregating Lac⁻ cells. Explain how strain C cells probably form. Does strain C contain the *dna*⁻(Ts) allele? Could this phenomenon occur for an *E. coli* strain that is mutant in either of the genes *dnaA* or *dnaE*?

8-56   An *E. coli* strain is grown for five seconds in medium containing ³H-thymidine. The cells are chilled and broken open by ultrasonic vibration, a procedure which fragments the DNA to pieces whose molecular weight is approximately 5 million. If the broken cells are centrifuged, almost all of the radioactivity is at the bottom of the tube. If the extract is treated with ether or acetone, none of the radioactivity is at the bottom—all is in the supernatant. If the cells are labeled for five seconds with ³H-thymidine and then grown for one minute in nonradioactive thymidine, none of the radioactivity is at the tube bottom. Explain this observation.

8-57   If *E. coli* is labeled for a few seconds with ¹⁴C-methionine, radioactivity is incorporated into the DNA. What does this represent, and what function does it serve?

9

DNA Repair

● 9-1   Name four ways that the base sequence of the DNA of an organism can be changed.

● 9-2   Name the two most common changes in DNA that occur spontaneously without attack by external agents.

● 9-3   Name agents that
(a) Cause production of thymine dimers.
(b) Cause production of 5,6-dihydroxydihydrothymine.
(c) Increase the rate of depurination.
(d) Cause interstrand cross-links.
(e) Cause single-strand breakage.
(f) Cause double-strand breakage.

9-4   The survival curve assay for radiation-induced killing is based on the inability of damaged cells to form a visible colony. One usually thinks of the damage that inhibits DNA replication but clearly there are other possibilities.
(a) Name several other possibilities.
(b) Survival curves are often single-hit—that is, only one element in the cell needs to be damaged. Is such first-order

inactivation kinetics compatible with any of your suggestions in part (a)?

●9-5 Which of the following is a true statement about DNA repair?
(a) Thymine dimers are usually formed between adjacent thymines in the same strand.
(b) Thymine dimers are usually formed between two thymines in different strands.
(c) The photoreactivation enzyme cleaves all thymine dimers in an ultraviolet-irradiated cell.

9-6 What is the purpose of light in the functioning of the photo-reactivating enzyme? What common biochemical reaction is probably replaced by the light?

9-7 A cell possesses a repair system capable of removing half of the damage produced by some agent. Which of the following statements might be true of a single-hit survival curve and of a multi-hit survival curve?
(1) A dose yielding x percent survival when repair does not occur would yield $2x$ percent survival if repair occurs.
(2) The dose required for a particular percentage of survival in the absence of repair is doubled when repair occurs.

9-8 Suppose 30 years ago you had observed that DNA replication in bacteria is inhibited by radiation from a germicidal lamp that emits ultraviolet light. Clearly the damage could result from direct absorption of the ultraviolet light by the DNA, but it was necessary to consider the possibility that some other molecule absorbed the light and afterwards damaged the DNA. Without using modern physical and chemical techniques what simple spectral measurement might you have done to investigate this question?

●9-9 Name the repair system that
(a) Cleaves thymine dimers.
(b) Cleaves N-glycosylic bonds.
(c) Cleaves phosphodiester bonds.

9-10 Cells having a very low concentration of either polymerase I or DNA ligase are somewhat more sensitive to ultraviolet light than wild-type cells but not as sensitive as Uvr- mutants. How is this fact compatible with the fact that both polymerase I and ligase are needed in excision repair?

9-11   The ability of ultraviolet-irradiated T4 phage to form plaques is the same on both Uvr⁺ and Uvr⁻ bacteria. How might you explain this fact?

● 9-12   Name two features of SOS repair that distinguish it from all other repair systems.

9-13   A bacterial culture in which the DNA is labeled with ³H-thymidine is irradiated with ultraviolet light. If, afterwards, the cells are exposed to visible light for a long time, a maximum of 30 percent of the dimers are cleaved by photoreactivation (PR). If, instead, after irradiation they are incubated in the dark for a long time in a dilute buffer (in order to prevent cell division), 40 percent of the dimers are cleaved. This is called dark repair (DR). In order to determine whether the set of dimers cleaved by PR overlaps the set cleaved by DR, experiments are done in which the two repair systems are allowed to act sequentially and the percent of thymine dimers remaining is measured. Listed below are the results of four possible but different experiments. In each, what conclusion could you draw about the properties of the two repair systems?

	Sequence	Percent remaining dimers
(a)	PR → DR	30
(b)	PR → DR	45
(c)	DR → PR	45
(d)	PR → DR or DR → PR	40 (by both sequences)

9-14   Some years ago, a mutant called B/r was isolated from a population of *E. coli* heavily irradiated with ultraviolet light. This mutant is very resistant to ultraviolet light when compared to the parent strain B. Whereas the isolation of such a mutant from the population of surviving cells is not surprising, it is more difficult to understand why among the survivors of a similar irradiated culture a highly radiosensitive strain known as B$_S$ could have been found. Propose a mechanism to explain this latter discovery. Would you expect the original colonies of both B/r and B$_{S}$, to have been pure or to have contained parental wild-type cells?

9-15   An unexcised thymine dimer produces a partial block to DNA replication. It is observed that if there are unexcised thymine dimers in parental strand, the daughter strands contain large gaps which are frequently several thousand nucleotides long. Would you expect to find gaps in both daughter strands if there were only a single thymine dimer in the parent molecule? If not, in which strand? Explain.

9-16   In an unirradiated cell, if RNA synthesis is blocked by rif-ampicin (an inhibitor of RNA polymerase), DNA molecules in the act of replication can be completely replicated, but a second round of replication cannot begin. Would this also be true for a heavily ultraviolet-irradiated cell?

9-17   A large dose of ultraviolet irradiation can kill an appreciable fraction of a population of wild-type cells even if the dose is not large enough to saturate the repair apparatus. Explain this phenomenon.

9-18   A bacterial repair system called X removes thymine dimers. You have in your bacterial collection the wild-type (X^+) and an X^- mutant. Phage λ, when ultraviolet-irradiated and then plated, gives a larger number of plaques on X^+ than on X^- bacteria. It has been proposed on the basis of survival curve analysis that the X enzyme is inducible. To test this proposal, ultraviolet-irradiated λ phage are adsorbed to both X^+ and X^- bacteria in the presence of the antibiotic chloramphenicol (which inhibits protein synthesis). No thymine dimers are removed in the X^- cell and 50 percent are removed in the X^+ cell. In the absence of chloramphenicol, the same results were obtained.
(a) Is X an inducible system?
(b) Suppose 5 percent of the thymine dimers are removed in the presence of chloramphenicol and 50 percent in its absence; how would your conclusion be changed?

9-19   It is possible to induce a mutation in $polA^-$ strains with smaller doses of ultraviolet radiation than are needed to induce a mutation in wild-type $E. coli$ strains. Why? Would you expect the same type of result if you compared the dose of ultraviolet radiation required to induce a mutation in $uvrA^-$ cells with that required for wild-type cells?

10

Mutations, Mutants, and Mutagenesis

● 10-1 Define the terms mutant, mutation, mutagen, and mutagenesis.

●10-2 Distinguish the symbols Gal⁺ and *gal⁺*.

10-3 A replication error is an example of what type of mutagenesis?

● 10-4 Distinguish the terms nonsense mutation and missense mutation.

10-5 One variety of a temperature-sensitive mutation is the cold-sensitive (Cs) mutation, which has a mutant phenotype below a particular temperature. Table 10-5 describes several mutations

TABLE 10-5

	32°C	*37°C*	*42°C*
ess2 (Ts)	+	−	−
ess5 (Ts)	−	−	+

in an essential gene *ess*, in which + and − refer to colony formation and the lack thereof, respectively. What would be the phenotype of an *ess2*(Ts)*ess5*(Cs) double mutant?

10-6   Consider a DNA molecule with the structure shown in Figure 10-6.
(a) If during replication, the T at position 2 is read as an A, what are the results of the first and second rounds of replication?
(b) If the replication apparatus copies the AT pair at 2, and then copies the AT at 5, skipping 3 and 4, what are the results of the first and second rounds of replication?

FIGURE 10-6

A	T	G	C	A	T	T	A	T	G	C	C
T	A	C	G	T	A	A	T	A	C	G	G

        1   2   3   4   5   6

•10-7 Explain why some base-pair changes in DNA fail to give rise to a protein with an amino acid substitution.

10-8   Why does the substitution of a single amino acid sometimes fail to produce a protein with reduced activity (that is, a mutant protein)? What kinds of amino acid changes would most likely result in a mutant protein?

10-9   Often dyes are incorporated into agar in order to determine whether a bacterium can utilize a particular sugar as a carbon source. For instance, in EMB agar containing a sugar X, a bacterium which has phenotype X^+ makes a purple colony and one which has phenotype X^- makes a pink colony. This happens because fermentation of the sugar produces acid, which changes the dye to a purple color. On EMB agar containing lactose, xylose, and maltose, what would be the color of the colonies produced by bacteria having the following phenotypes? (1) $Lac^+Xyl^+Mal^+$, (2) $Lac^-Xyl^-Mal^+$,   (3) $Lac^+Xyl^-Mal^+$,   (4) $Lac^-Xyl^+Mal^-$,   (5) $Lac^-Xyl^-Mal^-$?

10-10 Refer to question 10-9. If a population of Lac^+ cells is treated with a mutagen that produces Lac^- mutants and the population is allowed to grow for many generations, a few pink colonies will be found on EMB agar plates among a large number of purple ones. If the mutagenized population is plated on the agar immediately after the population has been treated with the mutagen, some colonies appear which are termed sectored—they are purple on one side and pink on the other side. Explain the color distribution of these colonies.

10-11 Some antibiotics, e.g., sulfonilamide, inhibit bacteria reversibly in the sense that growth of inhibited bacteria is restored if the antibiotic is removed from the growth medium. Suggest a method for isolating a Sul⁻ mutant from a population of Sul⁺ bacteria.

10-12 *E. coli* polymerase I possesses several enzymatic activities. Two important activities are the polymerizing function and the $3' \rightarrow 5'$ exonuclease. Mutant polymerases have been found which either increase or decrease mutation rates in an organism containing the mutant enzyme. A mutant that increases the mutation rate is called a mutator; a mutant that decreases the mutation rate is called an antimutator. The mutator and antimutator activities are usually a result of changes in the ratio of the two enzymatic activities described above. How do you think the ratios change in a mutator and in an antimutator?

10-13 Define the term silent mutation.

10-14 S1 nuclease makes a double-strand break at the site of two unpaired bases (a base-pair mismatch). It has been observed that if DNA molecules isolated from a strain of λ phage maintained in a laboratory in the United States are renatured with DNA molecules of a phage that is presumably genetically identical to the first strain but obtained from a laboratory in France, approximately half of the renatured DNA molecules are broken by S1 nuclease. Explain this observation.

10-15 Most amino acid-requiring bacterial mutants require a concentration of the amino acid of at least 20 µg/ml in the agar to achieve a normal colony size (3 mm diameter). A strain of *E. coli* is observed to produce colonies only 1 mm in diameter on a minimal agar yet colonies of normal size if the agar is supplemented with leucine at a concentration of 8 µg/ml. Explain.

10-16 Several hundred independent missense mutants, altered in the A protein of tryptophan synthetase, have been collected. Originally, it was hoped that at least one mutant for each of the

186 amino-acid positions in the protein would be found. However, fewer than 30 of the positions were represented with one or more mutants. Suggest some possibilities to explain why this set of missense mutants was so limited.

10-17   Suppose a bacterium is treated with a chemical that alters a particular adenine at a single site in a gene. Because of this alteration, the adenine appears to the DNA and RNA polymerases to be a guanine, and after several rounds of replication of the DNA there will be a DNA molecule with a GC pair at the site where there was an AT pair. Let us consider the bacterium before any DNA replication occurs. Two requirements must be met at that time if the bacterium is to lose the ability to make a functional gene product from that gene. What are they?

● 10-18   Which of the following amino acid substitutions would surely yield a mutant phenotype?   (1) Pro to His; (2) Lys to Arg; (3) Ile to Thr; (4) Ile to Val; (5) Ala to Gly; (6) Phe to Leu; (7) Tyr to His; (8) Arg to Ser.

10-19   Define the terms transition and transversion.

10-20   When studying auxotrophic bacterial mutants induced by the mutagen NNG, it is often observed that the colonies on agar are smaller than wild-type colonies even when there is an optimal supply of exogenous nutrient in the agar. Explain why the colonies are small.

10-21   Explain the basis of mutation by the following mutagens: 5-bromouracil, 2-aminopurine, nitrous acid, and acridine orange. Also, state whether they induce transitions (for example, an AT pair replaced by a GC pair or *vice versa*), transversions (for example, an AT pair replaced by a TA pair), or frameshifts.

10-22   Spontaneous mutations (those which arise without requiring exposure to a mutagen) are very frequenty frameshift mutations. Assuming that the base-recognition systems are not always perfect, explain how a frameshift mutation could arise by an error in either replication or recombination.

10-23   A phage T4 mutant in the *rII* gene is isolated. A revertant of this mutant is isolated that has the wild-type (*r+*) phenotype. If this revertant is crossed with authentic *r+* phage and the phage progeny are plated, many plaques are found with the mutant phenotype. Phage stocks are prepared from a large number of these mutant plaques and the stocks are systematically crossed against one another. The plaques are found to fall into two classes A and B defined as follows. If a phage in class A is crossed with a phage in class B, recombinants with an r^+ phenotype are produced. However in a cross between two phage from the same class, no recombinants with an r^+ phenotype are found. Furthermore, it is found that if the phage in either class A or B are crossed with the original mutant, r^+ recombinants are found only in a cross with a class B phage. Suppose it were known that the original mutation was a deletion of two bases; what kind of mutation or mutations would be present in the class B phage?

10-24   Suppose you isolated a mutant bacteriophage that was capable of forming a plaque on bacterial strain A but not on strain B. How would you go about deciding whether you had a single point mutant, a double mutant, or a deletion? First, answer the question assuming that it is the only mutant that has ever been isolated in this phage; then, answer it assuming that you have a very large collection of mutants which have already been mapped.

10-25   One hundred Lac⁻ mutants are examined separately to measure reversion frequencies. Of these, 45 revert at a frequency of 10^{-5}; 49 at 3×10^{-6}; 3 at 3×10^{-11}; and 3 at 10^{-10}. What type of mutants are contained in the class whose reversion frequency is 10^{-10}—single point mutants, amber mutants, double mutants or deletions?

10-26   How would you identify genetically whether a particular mutant is a frameshift, missense, or nonsense mutant? If it is a nonsense mutant, how would you tell if it is an amber, an ochre, or an opal mutant?

10-27   Consider a bacterial gene containing 1000 base pairs. As a result of treatment of a bacterial culture with a mutagen, mutations in this gene are recovered at a frequency of one mutant per 10^5 cells. One of these mutants is taken and grown and a pure culture of this mutant is obtained. This culture is then treated with the same mutagen and revertants are found at a frequency of one per 10^5 cells. Would you expect the gene product obtained from the revertant to have the same amino-acid sequence as the wild-type cell? Explain.

10-28   An amber mutation (his1) and a frameshift mutation (his2) are known to map close together within the same cistron. Each mutant reverts at a frequency of 1 revertant per 10^6 cells. A double mutant has been constructed which carries both mutations in the *his* gene.

(a) What would you expect for the frequency of His$^+$ revertants from this strain?

(b) How might you account for a reversion frequency of one His$^+$ revertant per 10^7 cells, if such an observation were made?

10-29   An enzyme contains 156 amino acids. (This number has no significance for the problem.) Suppose amino acid 28, which is glutamic acid, is replaced by asparagine in a mutant, and, as a result, all activity is lost. In this mutant protein, amino acid 76, which is asparagine, is replaced by glutamic acid and full activity of the enzyme is restored. What can you say about amino acids 28 and 76 in the normal protein?

10-30   Biological membranes are composed mainly of lipids and protein. About one-third of the dry mass of the *E. coli* cell membrane is lipid and all of this is phospholipid. One approach to delineating the roles of membrane phospholipids is through the isolation of mutants defective in the biosynthesis of various species of membrane phospholipids. Describe a procedure for selection of *E. coli* mutants which are temperature-sensitive for synthesis of cell-membrane phospholipids. Use L-glycerol 3-phosphate (GP) as a phospholipid precursor. As a starting strain, you are given an *E. coli* strain that is constitutive for uptake of GP and that cannot ferment GP, owing to mutations which destroy the activities of alkaline phosphatase and GP dehydrogenase. (Hint: A valuable trick is to use the fact that cells containing large amounts of radioactivity will often die by radioactive decay if stored for several days or weeks.)

10-31   For many genes there is no convenient means to enrich a culture for mutants so the mutant fraction is annoyingly small. The following technique has proved to be valuable to isolate such mutants. If a mutation in gene *b* is needed, a culture of a strain that has a mutation in a nearby gene (say, *a*) is treated with the mutagen nitrosoguanidine (most other mutagens are not effective in this technique) and revertants of the *a* mutant are selected. Many of these revertants are also *b* mutants. What is the principle underlying this technique?

10-32   A mutagen MTG has the property that when a bacterial culture is exposed to it, double mutants that are very near one another are frequently produced. Distant double mutants are produced at very low frequency as would be true of an ordinary mutagen. The data shown in Figure 10-32A can also be obtained with any strain of *E. coli* that requires an amino acid and that can take up ³H-thymidine (a specific label for DNA). The dashed line shows the data obtained if the required amino acid had not been removed.

A bacterium requiring nutrients A, B, C, ... is starved for a required amino acid for two generation times. Then the amino acid is restored and MTG is added in two-minute pulses (ample time for mutagenesis). When the culture is tested for reversion, the reversion frequencies are found to have increased about 100-fold; however, for any particular pulse only one gene shows the higher reversion frequency. Furthermore, the temporal sequence of the increases in reversion frequency follows map order starting from an arbitrary point in the circular genetic map of *E. coli*. That is, if the map is circular, as shown in Figure 10-32B, the time at which each gene A, B, C, ... shows the increase in reversion frequency can be plotted as in Figure 10-32C.

(a) What special property does MTG probably have?

(b) What does amino acid starvation do to the DNA replication cycle?

(c) What is the effect of re-addition of amino acids?

(d) What do these experiments tell about DNA replication in *E. coli*?

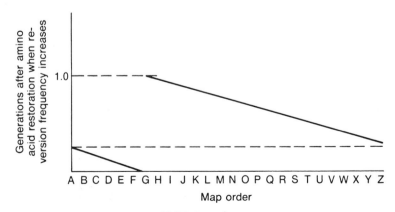

FIGURE 10-32A

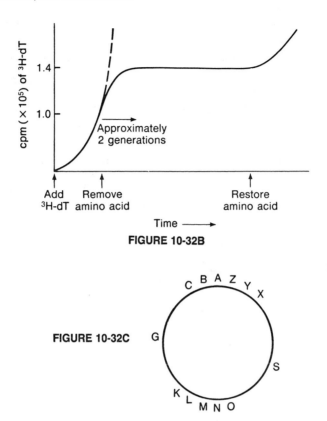

FIGURE 10-32B

FIGURE 10-32C

10-33 Among your collection of Gal⁻ mutants, you have heat–sensitive mutants (Gal⁻ only at high temperature), cold-sensitive mutants (Gal⁻ only at low temperature), and nonsense (chain termination) mutants. The distribution of these mutants among genes is as shown in Table 10-33. From these data, estimate the relative size of the three genes (assume the mutation rate per base pair is constant).

TABLE 10-33

Class of mutation	Distribution		
	galA	galB	galC
Nonsense	20	50	100
Heat-sensitive	5	100	5
Cold-sensitive	1	0	20

10-34   Assume that Figure 10-34 is an idealized map of the *rII* region of phage T4. Several point mutations are indicated by solid circles. A new *rII* mutation *z* is found that does not complement any of the above mutations, and it does not revert. In genetic crosses, *z* recombines with any of the above mutations and produces wild-type phage. What explanation can you suggest for this observation?

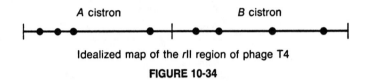

Idealized map of the *r*II region of phage T4

FIGURE 10-34

11

Transcription

•11-1 (a) From what substrates is RNA made?
(b) On what template?
(c) With what enzyme?
(d) Is a primer required?

•11-2 (a) Write down the two RNA sequences which could conceivably result from complete transcription of the following DNA duplex.

$$5' \; A \; G \; C \; T \; G \; C \; A \; A \; T \; G \; \; 3'$$
$$3' \; T \; C \; G \; A \; C \; G \; T \; T \; A \; C \; \; 5'$$

Indicate the 5' and 3' ends of each transcript.
(b) In an *in vitro* system would both transcripts be made?

•11-3 Describe the differences, if any, between the chemical reactions catalyzed by DNA polymerase and RNA polymerase.

•11-4 Which of the following events occur during RNA synthesis? (1) Binding of DNA polymerase to DNA; (b) binding of RNA polymerase to DNA; (3) binding of a σ factor to RNA polymerase; (d) binding of a σ factor to DNA polymerase.

●11-5   What chemical groups are present at the origin and terminus of a molecule of mRNA that has just been synthesized?

●11-6   Define the following terms: RNA polymerase core enzyme; σ factor; RNA polymerase holoenzyme; promoter; operator.

11-7   It is very rare that both DNA strands in any particular region serve as a template for RNA synthesis. Describe an old observation concerning mutations in fungi from which the above result might have been predicted.

11-8   When the RNA polymerase reaction is carried out *in vitro*, it is found that at physiological ionic strength fewer regions of a particular DNA molecule are transcribed than when the ionic strength is very low. Furthermore, at low ionic strength, the regions described differ markedly from those observed to be transcribed *in vivo*. Explain.

●11-9   What are the functions of the core enzyme and the holoenzyme *in vivo* ?

11-10   (a) How does the DNase protection method work and what information does it yield?
        (b) What information is provided by the dimethyl sulfate protection method?

●11-11 (a) What is mRNA?
        (b) How does mRNA sometimes differ from a primary transcript?
        (c) Define coding strand and antisense strand.
        (d) Define cistron and polycistronic mRNA.
        (e) What parts of a mRNA molecule are not translated?

●11-12 What is a transcription unit and is it the same thing as a gene?

11-13   Describe two types of promoter mutations.

●11-14 What is the difference between an open-promoter complex and a closed-promoter complex?

11-15   The rate of initiation of new RNA strands is greater with a superhelical template than with a linear DNA molecule or a nicked circle. Propose an explanation for this phenomenon.

11-16   Some DNA molecules have complementary single–stranded ends which allow circularization by means of base pairing of these segments. DNA ligase can be used to seal the single-strand breaks, yielding a covalent circle without superhelical turns. If RNA polymerase is bound to the DNA before the joining of the end segments and ligation and then the RNA polymerase is removed, the DNA circle has superhelical turns. What does this say about the binding of RNA polymerase to DNA?

11-17   How do the modes of action of the following antibiotics differ? Actinomycin D; rifampicin; and streptolydigin.

11-18   Assuming that you do not have available any data on the base sequence of any mRNA molecule, what information is available that suggests all promoters do not have the same base sequence?

11-19   How does one determine the first base transcribed in a promoter sequence?

●11-20   What is meant by the terms "upstream" and "downstream"?

11-21   The Pribnow box sequence is an example of what is termed a "consensus sequence"—namely, a base sequence from which other sequences having similar functions can be obtained by changing only one or two bases.
(a) What is the evolutionary significance of such a sequence?
(b) What is the biochemical significance of a conserved base, such as the conserved T in the Pribnow box?
(c) What biochemical differences might you expect between sequences that differ only slightly in the nonconserved bases?

●11-22   RNA polymerase lacks the nuclease activities of DNA polymerase. Name two activities possessed by RNA polymerase and not by DNA polymerase.

11-23   Why is the ssb protein not needed by RNA polymerase?

11-24   If the purified holoenzyme of *E. coli* RNA polymerase (the holoenzyme includes the σ subunit) is added to DNA in the presence of the four necessary nucleoside triphosphates, RNA is made. Analysis of the size of the RNA chains made shows mRNA molecules having a wide range of sizes, and the size distribution is nearly continuous. If a small amount of a cell extract (demonstrated to be free of ribonuclease activity) is added, smaller fragments of discrete size are observed. What is probably present in the extract?

●11-25   What features are present at the terminus of all bacterial transcripts?

●11-26   What is meant by a pause?

11-27   Radioactive (^{14}C) uridine is added to a growing culture of a bacterium for 10 seconds and then rifampicin is added. At two-minute intervals one-ml samples are removed and placed in an ice bath. The bacteria are lysed and DNase is added. After a digestion time of an hour to ensure that DNase digestion is complete, each sample is divided into two portions. One portion is precipitated and the other is hybridized with the DNA of a phage carrying the bacterial *cla* gene. The data are shown in Table 11-27. Plot the data on semilog paper and assume that mRNA decay is first-order.
(a) What fraction of the labeled RNA is stable RNA (rRNA and tRNA)?
(b) What is the half-life of the total bacterial mRNA?
(c) What is the half-life of *cla* mRNA?
(d) What was the purpose of the DNase treatment?

TABLE 11-27

Time in minutes	Acid precipitable radioactivity, cpm	Hybridizable radioactivity, cpm
t = 0	50,240	1820
t = 2	35,050	1375
t = 4	27,510	1060
t = 6	23,725	918
t = 8	21,850	720
t = 10	20,920	561
t = 12	20,460	431
t = 14	20,230	310
t = 16	20,110	248
t = 18	20,010	210

11-28 An RNA molecule is isolated having a 3'-OH terminus and a 5'-P terminus. What information does this fact provide?

11-29 What are the three steps in the production of a tRNA molecule from a primary transcript?

●11-30 In what ways are the templates for transcription in prokaryotes and eukaryotes different?

11-31 What RNA polymerases exist in eukaryotic cells? What are their locations and their functions *in vivo*?

●11-32 These questions refer to eukaryotic RNA.
(a) What is a cap?
(b) At which end of mRNA is the poly(A)?
(c) Are there eukaryotic mRNA molecules that do not contain either feature?

●11-33 (a) In what way, relating to polycistronic mRNA, do eukaryotic and prokaryotic protein synthesis differ?
(b) What is currently thought to be the mechanistic basis of that difference?

11-34 What is meant by heterogeneous nuclear RNA?

●11-35 (a) What are intervening sequences or introns?
(b) What is mRNA "splicing"?

11-36 A plasmid is prepared by genetic engineering to contain a gene that is a copy of the mRNA encoding the product of a particular gene. The plasmid DNA is isolated, denatured, hybridized with the

FIGURE 11-36

primary transcript of the gene. The electron micrograph drawn schematically in Figure 11-36 is obtained. (Thin and thick lines refer to single- and double-stranded nucleic acid respectively.) How many introns are there in the gene?

11-37   Modern genetic engineering techniques allow the movement of genes from one organism to another. Placing eukaryotic genes into *E. coli* has been of special interest.

(a) It has been observed that transferred animal genes are not always transcribed in *E. coli* yet transcription occurs much more frequently in yeast. Why?

(b) The animal genes can sometimes be connected to an *E. coli* promoter. Transcription of the gene then occurs but now the protein is not always synthesized. Again protein production is more efficient in yeast. Why?

12

Translation: The Information Problem

●12-1 A DNA molecule has the structure

T	A	C	G	G	G	A	A	T	T	A	G	A	G	T	C	G	C	A	G	G	A	T C
A	T	G	C	C	C	T	T	A	A	T	C	T	C	A	G	C	G	T	C	C	T	A G

The upper strand is the coding strand and is transcribed from left to right. What is the amino acid sequence of the protein encoded in this DNA molecule?

12-2   Which of the following amino acid changes can result from a single base change?
(a) Methionine to arginine.
(b) Histidine to glutamic acid.
(c) Glycine to alanine.
(d) Proline to alanine.
(e) Tyrosine to valine.

12-3   Which amino acids can replace arginine by a single base-pair change?

12-4   Nitrous acid HNO_2 induces mutations by deaminating cytosine and adenine, thereby converting them respectively to uracil and hypoxanthine (which pairs like guanine). Which of the following amino acid changes could be induced by nitrous acid? Arg to His; Leu to Ser; Tyr to His; Met to Ile; Gly to Asp.

12-5   Ribonuclease contains 124 amino acids. What is the least number of nucleotides you would expect to find in the gene encoding the protein? The first amino acid is not methionine.

12-6   The amber codon UAG does not correspond to any amino acid. Some strains carry suppressors which are tRNA molecules mutated in the anticodon, enabling an amino acid to be placed in at a UAG site. Assuming that the anticodon of the suppressor differs by only one base from the original anticodon, which amino acids could be inserted at a UAG site? At a UAA site?

12-7   Suppose that in another world the code is a doublet code having four bases. What is the maximum number of amino acids that could be in proteins?

12-8   There are several arginine codons. Suppose you had a protein that contained only three arginines (Arg-1, Arg-2, Arg-3). In a particular mutant, Arg-1 is replaced by glycine. In another mutant, Arg-2 is replaced by methionine. In still another mutant, Arg-3 is replaced by isoleucine. Suppose several hundred other mutants at various sites are isolated. Which other amino acids would you expect to find replacing Arg-1, Arg-2, and Arg-3, assuming only single-base changes?

12-9   *E. coli* DNA has a molecular weight of 2.7×10^9. If all this nucleotide sequence coded for proteins, how many proteins of average size would that correspond to? (An average protein has a molecular weight of 50,000 and the amino acid residues have an average molecular weight of 110).

•12-10   The nucleotide sequence around the initiation site on the message for the coat protein of the RNA phage R17 is

5'-..G A A G C A U G G C U U C U A A C U U U..-3'

(a) Which codon specifies the first amino acid of the protein?
(b) Why does the trinucleotide sequence, UAA, toward the right end of the sequence not cause chain termination?

12-11   You have isolated an acridine-induced mutant of a particular
         protein and have purified the mutant protein. If the wild-type
         and mutant proteins are cleaved with trypsin and the peptides are
         examined, it is found that only one peptide is altered.

         (a) Where is this peptide located in the protein?
         (b) The peptide from the wild-type protein is Leu–
         Met-Ser-Val-Glu; that from the mutant is Leu-Thr-Glu-Arg. How
         many extra bases have been added to the DNA by growth in the
         presence of acridine orange? Identify and locate the new base or
         bases in the base sequence.

12-12   Which base–substitution mutagens can cause a reversion to the
         original amino acid by replication of a mutant DNA in the
         presence of the same mutagen that induced the original
         mutation?

12-13   Suppose that you are making use of the alternating copolymer
         GUGUGUGUGU... as an mRNA in an *in vitro* protein-synthe-
         sizing system. Assuming that an AUG start codon is not needed in
         the *in vitro* system, what peptides are made by this mRNA?

12-14   Suppose you are synthesizing a random copolymer of G and U using
         equimolar quantities of G and U nucleotides. If this polymer is
         used as a mRNA, which amino acids can be incorporated into
         protein and at what relative frequencies?

12-15   For an *in vitro* protein synthesis experiment to study the genetic
         code, a synthetic mRNA species was synthesized with
         polynucleotide phosphorylase, using adenosine diphosphate (ADP)
         and cytidine diphosphate (CDP) in the ratio 5:1. The base
         sequence produced in this synthesis is random. The following
         amino acids were incorporated into polypeptide material in the
         following relative amounts: lysine, 100; asparagine, 20;
         glutamine, 20; threonine, 24; proline, 4.8; histidine, 4. Using
         this information, identify possible codons for the above amino
         acids and give their frequencies.

●12-16   Which amino acids can be incorporated into a protein if the
         following repeating polymers are used as mRNA molecules? Assume a
         special start codon is not needed *in vitro*.
         (a) C G A C G A C G A C G A

(b) A U G A U G A U G A U G ...
(c) A U A A U A A U A A U A ...

12-17 Write down the codons which, if altered by a single base change, could give rise to chain termination.

12-18 Under what circumstances might the appearance of a chain termination codon not result in loss of activity of a protein?

12-19 Suppose you are testing the idea that for each segment of DNA, both strands of DNA are used as a template for synthesis of mRNA. Which of the following results could be taken as evidence to support this idea?
(a) Mutations occur in pairs in that frequently two amino acids in a mutant protein are changed.
(b) Mutations are paired in that frequently two different proteins are simultaneously affected.

12-20 Assuming the hypothesis in 12-19 is correct and assuming that chain termination occurs in the standard way using only the known chain termination mutations, would two proteins always, never, or sometimes be prematurely terminated by the same change in a base pair? Explain.

12-21 The first codon that was identified was UUU for phenylalanine. The insolubility of polyphenylalanine allowed several other codons to be identified by a simple trick. Each nucleotide X was separately coupled to the 3'-OH terminus of polyuridylic acid (poly(U)). The coupling was heterogeneous in that some poly(U) molecules terminated with one X moiety, some with two, some with three, etc. However, the addition of a single X was more probable than two, and two more probable than three, etc.
(a) Which amino acids, and in what ratio, will be at the terminus of polyphenylalanine when each of the four bases is separately coupled to poly(U)?
(b) If each of the sixteen dinucleotides could be coupled to poly(U), could the entire code be worked out? Explain.

12-22 What is meant by wobble?

12-23 Considering both the code as we know it and the wobble hypothesis, what is the minimum number of tRNA molecules needed to recognize the 61 codons corresponding to amino acids?

12-24   (a) State various hypotheses to explain the fact that the start codon AUG frequently occurs within a protein but is not used as a start.
(b) What is the only circumstance under which a stop codon can occur within a coding sequence yet not cause termination? (Do not consider the possibility of a suppressor.)
(c) What signals an initiating AUG in prokaryotes?
(d) What is the signal in eukaryotes?

12-25   Human hemoglobin is a major oxygen-transporting protein and is the sole protein in red blood cells. Red blood cells have no nucleus or DNA; thus, hemoglobin is synthesized from a stable mRNA, whose 3'-terminal nucleotide sequence is

5'-A A G U A U C A C U A A G C-3'.

The carboxyl-terminal dipeptide of hemoglobin is -Tyr-His-COOH. A number of mutant hemoglobins from patients with genetic diseases exhibit elongated polypeptide chains. One such mutant hemoglobin ends in -Tyr-His-Leu-Ser-COOH. What is the mutation responsible for this elongated hemoglobin? Hint: Only one base is involved. There are no known nonsense suppressor mutations in human cells, so the mRNA for hemoglobin must itself be altered. Assume that protein synthesis can terminate at the 3' end of a molecule of mRNA as well as at a termination signal before the 3' end.

12-26   The diagram in Figure 12-26 represents a piece of DNA from the middle of an active wild-type gene. The hyphens indicate the reading frame of the corresponding mRNA. X and Y are two unknown bases; the prime indicates the complementary base.
(a) What is the sequence of amino acids in the indicated four residues? Label the amino and the carboxyl termini.
(b) The change of a single base pair will give rise to a nonsense codon from the above sequence. Give the possible base-pair changes that will give a nonsense codon.
(c) In a certain mutant you find the amino acid sequence Cys-Pro-Tyr-Met. Assuming the occurrence of only a simple mutational event (that is, one not involving several base pairs), how does the mutant DNA sequence differ from that of the

| 5' | X | ATA | TAG | GGG | GCA | Y | 3' | Strand 1 |
| 3' | X' | TAT | ATC | CCC | CGT | Y' | 5' | Strand 2 |

FIGURE 12-26

wild-type? What type of mutation is this?

(d) What must base X be, and why?

12-27   Consider a piece of DNA in a short nonessential (i.e., any amino acid sequence will function) part of a gene (Figure 12-27). Note that only the strand from which the messenger RNA is copied is shown. (The other strand is irrelevant to this problem.) Suppose that, by acridine orange treatment, a new sequence is created, so two G's are inserted at the point indicated. This will inactivate the protein because reading will be out of phase. However, if the G at the second indicated point is then mutated to become a C, an active protein (but not necessarily the same one) is made. Explain.

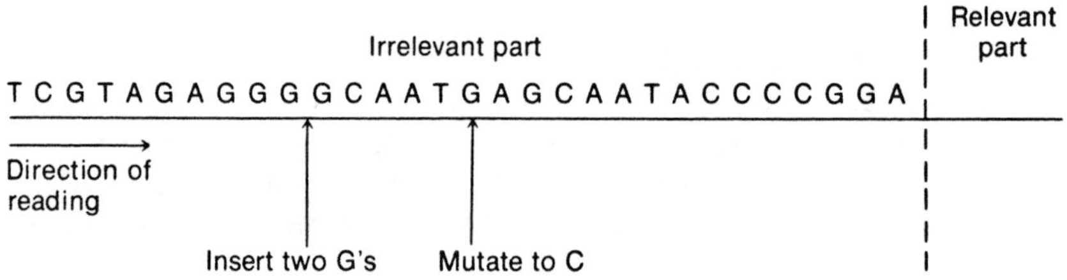

FIGURE 12-27

12-28   Consider two adjacent cistrons *A* and *B* on the same polycistronic mRNA.

(a) An acridine-induced mutant that prevents synthesis of protein *B* maps in gene *A* rather than in gene *B*. Growth of this mutant in a medium containing an acridine yields A^+ revertants that are also B^+. Explain these events.

(b) A B^+ revertant is isolated that also maps in gene *A*. This revertant is still A^-. Furthermore, growth in acridine can no longer induce formation of an A^+ mutant. If a suppressor is introduced into the B^+ revertant, the phenotype becomes A^-B^-. Explain what is happening.

●12-29   Which of the following properties are essential for the functioning of an aminoacyl synthetase? (1) Recognition of a codon; (b) recognition of the anticodon of a tRNA molecule; (3) recognition of the amino-acid recognition loop of a tRNA molecule; (4) ability to distinguish one amino acid from another.

12-30   Which of the following properties are essential for the function of a tRNA molecule? (1) Recognition of a codon; (2) recognition of an anticodon; (3) ability to distinguish one amino acid from another; (4) recognition of DNA molecules.

● 12-31   (a) What amino acid is bound to charged $tRNA^{Leu}$?
        (b) What amino acid is bound to seryl-$tRNA^{Leu}$?

12-32   Suppose a chemical were introduced into a cell that immediately converted to valine any isoleucine that had just been attached to isoleucyl-tRNA. Could protein synthesis occur and, if so, how would the proteins made differ from normal proteins?

12-33   What is the shape of tRNA molecules? Show where the amino acids and the anticodons are located.

12-34   What are meant by the stems, loops, arms, and tertiary base pairs of tRNA?

12-35   What are the three types of hydrogen bonds in tRNA?

12-36   The binding of DNA-binding proteins, such as Cro and CAP, to DNA and of aminoacyl synthetase to its cognate tRNA have a common geometric feature. What is it?

12-37   What three requirements must be met if an aminoacyl synthetase is to charge a tRNA molecule?

● 12-38   What is meant by a suppressor tRNA and how does such a molecule arise?

12-39   Since nonsense suppressors are mutant tRNA molecules, how does the cell survive loss of a needed tRNA by such a mutation?

12-40   An amber suppressor supplies an amino acid at the amber termination triplet UAG. It would seem as if, in a cell containing an amber suppressor, many necessary proteins might not be terminated, so that possession of this suppressor would be lethal. Why is this not the case?

12-41   Why are ochre (UAA) suppressors usually weak suppressors?

12-42   A missense suppressor substitutes a particular amino acid for a "wrong" amino acid in a mutant protein. Describe two ways that missense suppression could occur. Why are missense suppressors usually weak? If a missense suppressor substitutes amino acid Y for amino acid X, would you expect it to suppress all mutations in which X is the "wrong" amino acid in the mutant? Explain.

12-43   Consider two genes a and b. In growth medium Q, a mutation in either a or b is lethal (that is, the genotypes a^-b^+ and a^+b^- are lethal combinations). A genetic cross is made between one bacterium with genotype a^-b^+ and a second with genotype a^+b^-. It is found that there are two types of recombinants that can grow in Q, the a^+b^+ wild-type and (surprisingly) the a^-b^- double mutant—that is, a^- suppresses the b^- mutation and b^- suppresses the a^- mutation. Propose a molecular explanation for this phenomenon.

12-44   Suppressors were originally recognized by their ability to suppress a large number of mutants of many types. Frameshift suppressors differ in that usually only a small number of frameshift mutations are suppressed. A necessary but not sufficient property of the mutation is that it is a single base addition. It is known that these suppressors are tRNA molecules which have been altered. Propose a way in which a tRNA could be altered to become a frameshift suppressor. If you wanted to mutate *E. coli* to produce a frameshift suppressor, which mutagen would you use? Do you think a frameshift mutation consisting of a single base deletion could be suppressed? Explain.

12-45   Devise a scheme to isolate temperature-sensitive amber suppressors.

12-46   Explain the following two phenomena.
(a) A suppressed missense or nonsense mutation often has a mutant phenotype at 42°C but not at 30°C, even when it is known independently that the suppressor itself functions equally well at both temperatures.
(b) A mutant protein has amino acid B at a site normally occupied by A. If the strain also contains a missense suppressor, the protein has some activity. However it is known that the particular suppressor substitutes amino acid D for C and never substitutes A for B.

12-47   Generally, suppressor mutations for the UAA type of nonsense mutation also suppress the UAG type of nonsense mutation. Why do you think that this occurs? Suppose you decide to look into this problem more thoroughly by isolating a larger set of UAA suppressor mutations. After extensive mutagenesis, you isolate some rare "restricted-range" UAA-suppressor mutations which suppress UAA and not UAG. You now also find suppression of certain missense mutations. Propose an explanation in terms of tRNA structure. What amino acid do you predict will be replaced by the latter class of suppressor tRNA molecules?

12-48   In starting their work on the nature of the genetic code, Crick and collaborators were confident that the N-terminal end of the T4 *rIIB* protein is not required for phage growth, i.e., it is, in that sense, dispensable. One of the reasons that they drew that conclusion was the existence of a deletion, named *r1589*, which spans the N-terminal end of gene *rIIB* and the C-terminal third of gene *rIIA* and yields a fused rIIA-rIIB protein which has rIIB but not rIIA function. To judge from the genetic data and electron microscope-heteroduplex mapping, the *r1589* deletion is about 900 to 1100 nucleotides long. Following are four numbers which represent guesses regarding the exact length of the *r1589* deletion: 980, 984, 988, 1088. The base sequence of the DNA encoding the C-terminal end of the rIIA protein and the N-terminal end of the rIIB protein indicates that a fourteen-nucleotide spacer separates the genes. This information is sufficient to enable you to eliminate some of these proposed values. What values must be wrong? (To put it another way, what rule must the *r1589* deletion, and any other deletion that fuses the rIIA and rIIB polypeptide chains together, follow?)

12-49   The genes coding for the ten enzymes required in histidine biosynthesis constitute a single transcription unit in *E. coli*. They are clustered in the DNA as shown in Figure 12-49 (I), where each number represents a gene. (The numbers also indicate the sequence of the corresponding enzymes in the biosynthetic pathway, with 1 the catalyzing the first reaction, 2 the second, etc. Suppose that you isolate a mutant that somehow has undergone a perfect inversion of two genes by breaking and rejoining at the points shown by arrows, in such a way that the entire nucleotide sequences of genes *5, 7, 8,* and *9* are preserved, but the order now is that shown in sequence II.
(a) Would you expect this mutant to be His+ (normal phenotype) or His-?

(b) Explain how, if at all, the inversion might affect the transcription of gene *4*.

(c) Explain how, if at all, the inversion might affect the translation of gene *4* message.

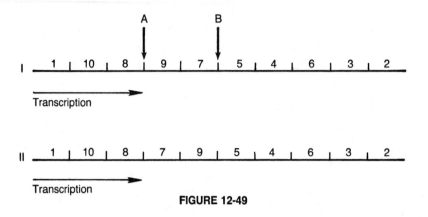

FIGURE 12-49

13

Translation: The Chemical Problem and Protein Synthesis

●13-1 What are the *s* values of prokaryotic and eukaryotic ribosomes, their subunits, and their RNA molecules?

●13-2 Do ribosomes have to be assembled with some special machinery or do they assemble spontaneously?

●13-3 What are the *L* and *S* proteins of ribosomes?

13-4 (a) What physical feature do tRNA and rRNA have in common? (b) A melting curve (optical absorbance vs. temperature) can be obtained for RNA just as for DNA. The melting curve for a mixture of purified 16S and 23S rRNA is virtually identical to that for 70S ribosomes (as long as one corrects the observed value of the absorbance for light scattering). What can you conclude from this observation?

13-5 Suppose you reconstitute a ribosome, omitting a particular ribosomal protein, and observe that tRNA will not bind to the reconstituted particle. Can you conclude that tRNA binds to that protein?

13-6   How would you distinguish 23S ribosomal RNA from phage RNA of the same size?

13-7   Which of the following are steps in protein synthesis in prokaryotes? (1) Binding of tRNA to a 30S particle; (2) binding of tRNA to a 70S ribosome; (3) coupling of an amino acid to a ribosome by an aminoacyl synthetase; (4) separation of the 70S ribosome to form 30S and 50S particles.

•13-8   Which of the following statements are true of tRNA molecules? (1) They are needed because amino acids cannot stick to mRNA. (2) They are much smaller than mRNA molecules. (3) They are synthesized without the need for intermediary mRNA. (4) They bind amino acids without the need of any enzyme. (5) They occasionally recognize a stop codon if the Rho factor is present.

•13-9   Which of the following are roles of tRNA in prokaryotes?

(1) To attach the amino acids to one another.

(2) To bring the amino acids to correct position with respect to one another.

(3) To increase the effective concentration of amino acids.

(4) To attach the mRNA to the ribosome.

•13-10   (a) With what chemical polarity are proteins synthesized on ribosomes?

(b) With what polarity is mRNA read?

(c) What nucleotide sequences in the mRNA signal the beginning and end of protein chains?

•13-11   The synthetic polynucleotide

$$5'-A\ G\ G\ U\ U\ A\ U\ A\ G\ G\ A\ A\ A\ A\ A-3'$$

is used as a mRNA in a system that does not require a start codon. What polypeptide is synthesized? Indicate the NH_2 and COOH termini.

13-12   For what purpose do 70S ribosomes dissociate during protein synthesis?

13-13   Excluding the enzymes needed to synthesize the amino acids, make a rough count of the number of known enzymes required to synthesize a single protein molecule. What fraction of *E. coli* enzymes is involved in protein synthesis?

13-14   Translation has evolved in a particular polarity with respect to the mRNA molecule. What would be the disadvantages of having the reverse polarity?

13-15   (a) A batch of reticulocytes actively making the protein hemoglobin were given radioactive amino acids for a brief period (where a "brief period" is a time short compared to the time required to polymerize the amino acids of a protein chain). Immediately thereafter, the soluble hemoglobin (that which is not attached to ribosomes) is examined for distribution of radioactivity in the chain. Which one of the following will be observed? (1) Label is found in amino acids at the amino end only. (2) Label is found in amino acids at all positions. (3) Label is found at the carboxyl end only.
(b) Suppose the reticulocytes were permitted to continue making hemoglobin with nonradioactive amino acids for a long time following the radioactive pulse described in (a). If the soluble hemoglobin were examined, which one of the following would be found? (1) Label at the amino end only. (2) Label in amino acids at all positions. (3) Label at the carboxyl end only.

13-16   What are the requirements for the union of one 30S and one 50S ribosomal subunit *in vivo* and *in vitro*?

13-17   Suppose you are studying protein synthesis *in vitro* with the synthetic polymer AUGUGUUAG. With your standard preparation of factors and ribosomes you find the expected production of free fMet-Cys. If you make the same standard factor preparation after infection by phage X19, you find no free dipeptide, but do find Val-Leu bound to the ribosome. Propose an explanation for this observation.

•13-18   Which of the following is the normal cause of chain termination? (1) The tRNA corresponding to a chain-termination triplet cannot bind an amino acid.
(2) There is no tRNA with an anticodon corresponding to a chain-termination triplet.
(3) Messenger RNA synthesis stops at a chain termination triplet.

13-19   In an *in vitro* protein-synthesizing system that requires an AUG

in the mRNA, all proteins start with formylmethionine. Consider a mRNA that has the sequence 5'-AUGAAAAAAAAA.
(a) What is the amino acid sequence of the protein that is made?
(b) If an antibiotic that inhibits translocation is added, what will be the product, if any, of the *in vitro* synthesis?

13-20   Which processes in protein synthesis require hydrolysis of GTP?

13-21   Explain the function of initiation, elongation, and termination factors.

13-22   How many high-energy phosphate-containing molecules are required to translate the mRNA of a protein molecule containing 100 amino acids? Assume that the protein-synthesizing system contains free amino acids, ribosomes, all necessary enzymes and factors, GTP, and ATP.

13-23   Suppose you have isolated a temperature-sensitive mutant of *E. coli*. At 42°C it is defective in protein synthesis. A culture growing at 30°C is warmed to 42°C at the same time ^{14}C-leucine is added. After 10 minutes at 42°C trichloracetic acid is added and the acid-insoluble material is isolated and examined; it is found to be very radioactive. If instead the cells are broken open and the cell extract is fractionated by centrifugation, it is found that all ^{14}C is in a fraction containing ribosomes; there is no free ^{14}C-labeled protein in this case. What process is probably temperature-sensitive?

13-24   Suppose you have a mutant of *E. coli* which grows normally at 30°C but slowly at 42°C. There is nothing obviously defective about the rates of DNA and RNA formation or protein synthesis shortly after the temperature is increased from 30°C to 42°C. In studying the induction of the synthesis of a particular enzyme with this mutant it is observed that if the inducer is removed, the rate of synthesis of the enzyme at 30°C decreases with a half-life of 5 minutes. At 42°C, however, the rate of synthesis does not decline but the amount synthesized per generation per cell decreases with a half-life of one generation. What process is probably inhibited at 42°C?

13-25   How does peptidyl transferase differ from a typical enzyme?

13-26   Apart from the difference in the number of factors needed for protein synthesis in prokaryotes and eukaryotes, what are the major differences in the modes of initiation?

13-27 Penicillin, tetracycline, erythromycin, chloromycetin, and puromycin are powerful inhibitors of bacterial growth. However, only the first three of these are used in the treatment of human diseases. Chloromycetin is used only in cases where all else fails. Explain these facts in terms of microbial and mammalian protein synthesis.

13-28 Can more than one ribosome read a single mRNA molecule concurrently?

13-29 Suppose polyribosomes did not exist and only a single ribosome could bind to a mRNA at any moment. Compared to a system having polyribosomes, will the number of protein molecules synthesized per unit time be greater, less, or the same? Explain..

13-30 What is the relation between the rate of codon production (in mRNA synthesis) and the rate of amino acid polymerization?

13-31 How long does it take to synthesize a typical protein (molecular weight 50,000) at 37°C?

13-32 Messenger RNA is usually measured by various hybridization techniques—that is, through the formation of a DNA-RNA hybrid by incubation of the mRNA with single-stranded DNA. In one technique, single-stranded DNA is adsorbed to a nitrocellulose filter, radioactive RNA is added, and then the mixture is incubated under conditions leading to hybridization. Free RNA does not bind to the filter, so the amount of radioactivity bound to a washed filter is a measure of the amount of RNA hybridized. In one experiment radioactive mRNA was isolated from a phage-infected cell one minute after infection and then hybridized with three different DNA molecules separately adsorbed to filters. The DNA molecules are shown in Figure 13-32, in which numbers refer to distance from the left end of the DNA, on a scale of 0 to 100. Wild-type phage (I) were used in the infection. The shaded regions in phage DNA molecules II and III represent bacterial DNA carried by the phage. Following hybridization with each type of DNA on two separate filters, one filter of each type was incubated with a ribonuclease (+RNase) that digests single-stranded RNA but not hybridized RNA; the other filter was untreated (−RNase). The data in Table 13-32 were obtained. From which region of the wild-type DNA was the mRNA transcribed?

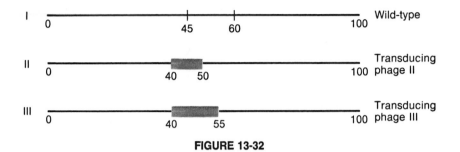

FIGURE 13-32

TABLE 13-32

	cpm on filter	
DNA on filter	−RNase	+RNase
I	1,250	1,245
II	1,260	820
III	1,242	418

14

Regulation of Prokaryotic Systems

●14-1 Define anabolic, catabolic, and metabolic.

●14-2 Distinguish negative and positive regulation.

14-3 What is meant by a partial diploid?

●14-4 What is the meaning of the following terms: repressed, induced, constitutive, coordinate?

●14-5 Is the regulation of gene activity in prokaryotes mainly transcriptional or posttranscriptional? At what steps of RNA synthesis is transcriptional regulation exerted in different systems?

14-6 Addition of lactose to a Lac$^+$ culture of *E. coli* stimulates transcription of the *lac* operon, but *allo*-lactose is the "true" inducer. From a molecular point of view, what is the distinction between an inducer (operationally speaking) and a true inducer? Also, what is meant by a gratuitous inducer?

14-7 Consider an operon having this order: repressor, operator, promoter, and cistrons a, b, and c. The repressor binds to the operator. When an inducer I is added, the repressor is inactivated and a single polycistronic mRNA is made. Normally, equal amounts of products A, B, and C are made; these products are needed for the utilization of I as a carbon source. A mutant is isolated which gives small colonies on an agar surface when I is the sole carbon source. Because you are interested in the proteins of so-called "leaky" mutants, you study this phenomenon further by measuring the activity of proteins A and B by standard biochemical tests in the presence and absence of I. You find that when I is present, the activity of A is 10 percent of normal and that of B is slightly higher than normal (normal means having the activity of the wild-type). Surprisingly, when I is absent, B activity is still found at near-normal levels, though A activity is absent. To study this further, you decide to try to map the mutation, perform dominance tests, check either the fingerprint or the amino acid sequence of A, and assay for C. A colleague tells you not to waste your time with this tedious work, since she can tell you the result of each test; furthermore, she is sure that the activity of the C protein will be nearly normal. What does your colleague predict?

14-8   The purpose of regulation of gene expression is to reduce wasted synthesis; that is, enzymes should be made only if needed. Mutants defective in regulation of a particular operon can be isolated (for example, those which have inactive repressors or inactive operators). Furthermore, we can break down the regulation by adding a gratuitous inducer, that is, one which is not a substrate of the induced enzyme. Consider an operon which is induced by the product X and makes an enzyme which allows X to be the sole carbon source.
(a) Suppose 10^6 wild-type and 10^6 repressorless bacteria are placed in a growth medium lacking X. In the repressorless cells 20 percent of the protein of the cell is Xase. If the bacteria are allowed to grow for 30 wild-type generation times (and the culture is diluted to prevent saturation), what will be the ratio of wild-type to repressorless cells at this time?
(b) Consider part (a) if X is present.
(c) Suppose the wild-type culture is grown for a very long time in the presence of a gratuitous inducer; what type of mutant might accumulate in the culture?

14-9   Would you expect the enzymes of the glycolytic pathway and the

Krebs cycle to be regulated? Consider repression versus other mechanisms.

14-10 Describe, in terms of mRNA synthesis, enzyme synthesis, and enzyme activity, what happens if lactose is added to a culture of a Lac$^+$ *E. coli* strain previously growing in a nutrient medium lacking all sugars. Assume that the amount of lactose added is consumed after two generations of growth.

14-11 Repeat 14-10 this time for a Lac$^+$ culture in lactose-containing medium when an amount of glucose is added that is consumed in one generation.

14-12 Which of the following conditions are ensured by the action of the cyclic AMP-CAP system? (1) That cyclic AMP levels do not become too high; (2) that enzymes are made when needed; (3) that enzymes are not made when not needed.

●14-13 Suppose *E. coli* is growing in a growth medium containing lactose as the sole source of carbon. The genotype is $i^-z^+y^+$. Glucose is then added. Which one of the following will happen?

  (1) Nothing.
  (2) Lactose will no longer be utilized by the cell.
  (3) *Lac* mRNA will no longer be made.
  (4) The repressor will bind to the operator.

14-14 A cell with genotype $i^+z^+y^+$ is in a growth medium containing neither glucose nor lactose (that is, it uses some other carbon source). How many proteins are bound to the DNA comprising the *lac* operon? How many if glucose is present?

14-15 A Lac$^+$ Hfr (that is, its genotype is $i^+o^+z^+y^+$) is mated with a female that is $i^-o^+z^-y^-$. In the absence of inducer, β-galactosidase is made for a short time after the Hfr and female cells have been mixed. Explain why it is made and why only for a short time. What would happen if the female were $i^+o^cz^-y^-$?

14-16 Explain how lactose molecules first enter an uninduced $i^+z^+y^+$ cell to induce synthesis of β-galactosidase.

● 14-17 For each of the following diploid genotypes, indicate first whether β-galactosidase can be made, and second, whether synthesis of β-galactosidase is inducible (I) or constitutive (C) and finally, whether or not each cell could grow with lactose as sole carbon source.

(a) $i^+z^-y^+/i^-z^+y^+$.

(b) $i^+z^+y^+/o^ci^+z^-y^+$.

(c) $i^+z^-y^+/o^ci^-z^+y^+$.

(d) $i^+z^+y^-/i^-z^-y^+$.

(e) $i^-z^+y^-/i^-z^+y^+$.

(f) $i^-z^+y^+/i^+o^cz^-y^+$.

(g) $i^+p^-z^+/i^-z^-$.

(h) $i^+o^cz^-y^+/i^+z^+y^-$.

(i) $i^+p^-o^cz^-y^+/i^+z^+y^-$.

(j) $i^-p^-o^cz^+y^+/i^-z^+y^-$.

14-18   For each of the *E. coli* diploids that follow, indicate whether the strain is inducible or constitutive, or negative for β-galactosidase and permease, respectively.

(a) $i^+o^+z^-+/i^+o^cz^+y^+$.

(b) $i^+o^+z^+y^+/i^-o^cz^+y^-$.

(c) $i^-o^+z^-y^+/i^-o^cz^+y^+$.

(d) $i^-o^+z^-y^+/i^+o^cz^+y^-$.

14-19   Suppose you have isolated a Lac⁻ mutant and by genetic analysis have found that the cell is z^+y^+; you have also found that the mutation, which you call i^*, is in the i gene. The diploid $i^*z^+y^+/i^-z^+y^+$ is constructed and found to be Lac⁻— that is, i^* is dominant. The diploid $i^*z^+y^+/o^ci^+z^+y^+$ is Lac⁺ (β-galactosidase is made). Suggest a property of the mutant repressor that yield this phenotype. Would $i^*z^+y^+/o^ci^+z^-y^+$ make β-galactosidase?

14-20   In order to study the regulation of the *lac* operon in *E. coli* you perform a diploid analysis with various regulatory and structural gene mutants which you have isolated. The results of your experiments are shown in Table 14-20. The numbers represent relative activity of the enzyme β-galactosidase. Your results are somewhat different from what you had expected.

(a) Complete the table: for experiments 5 through 10, write in the expected missing numbers.

(b) Compare experiments 1 and 4. Why do the values for induced β-galactosidase differ?

(c) Compare experiments 5 and 8 with experiment 1. Can you think of any explanation for the low values obtained in experiments 5 and 8 in the induced state?

(d) Examine experiments 9 and 10. Can you think of any explanations to account for these low values? Hint: The results in experiments 5 through 10 are different manifestations of the same phenomenon.

TABLE 14-20

Experi- ment	Genotype	Observed β-galactosidase activity		Expected β-galactosidase activity	
		Induced	Uninduced	Induced	Uninduced
1	$i^+o^+z^+$	100	0.1	100	0.1
2	$i^-o^cz^+$	100	100	100	100
3*	$i^-o^cz_1^-$	0.1	0.1	0.1	0.1
4	$i^+o^+z^+/i^+o^+z^+$	200	0.1	200	0.1
5	$i^-o^cz_1^-/i^+o^+z^+$	10	0.1	____	____
6	$i^+o^cz_1^-/i^-o^+z^+$	10	0.1	____	____
7	$i^-o^+z_1^-/i^+o^cz^+$	10	100	____	____
8	$i^+o^cz^+/i^-o^+z_1^-$	10	100	____	____
9	$i^-o^+z_2^-/i^+o^cz_3^-$	40	0.1	____	____
10	$i^-o^+z_3^-/i^+o^cz_2^-$	40	0.1	____	____

*z_1^-, z_2^-, and z_3^- are three different mutant alleles in the z gene. All three are missense mutations (a wrong amino acid is inserted).

14-21 A 30-base-pair double-stranded DNA segment containing the lac operator sequence has been inserted into a plasmid. The plasmid was introduced into a strain of E. coli having the genotype $i^+z^+y^+$. Once established, there were about thirty copies of the plasmid were present in each cell.
(a) The bacteria containing the plasmid were constitutive for β-galactosidase synthesis. Suggest an explanation.
(b) With continued growth of the culture, some inducible strains arose. What mutation probably produced inducibility?

14-22 An E. coli mutant is isolated that renders the cell simultaneously unable to utilize a large number of sugars, including lactose, xylose, and sorbitol. However, genetic analysis shows that each of the operons responsible for utilization of these sugars is free of mutation. What are the possible genotypes of this mutant?

14-23   What type of mutation of the *lac* repressor might be both constitutive and *trans*-dominant?

14-24   Although β-galactosidase and thiogalactoside transacetylase are coordinately regulated, their polypeptide chains are not produced in equimolar amounts. The following estimates have been given for messenger life-time and rate of translational initiation. Calculate the relative rates of production of the two polypeptide chains.

TABLE 14-24

Property	Time in seconds	
	β-galactosidase	Transacetylase
Half-life,	90	55
Average interval between translation starts on one mRNA	3	16

14-25   A phage containing a linear double-stranded RNA molecule and having a phage-encoded RNA replicase needed to replicate the RNA has just been described. Since this enzyme has many properties in common with *E. coli* DNA polymerase, a biochemist proposes to determine whether the RNA is also synthesized discontinuously, that is, in small fragments. The experiment seems simple enough, since synthesis of both RNA and DNA is shut off immediately after the host is infected. The following protocol, which is a standard procedure for making an analysis of fragments at a replication fork, is used; the host cell is infected; ^{3}H-uridine is added for two seconds; RNA is isolated and sedimented through an alkaline sucrose; the contents of the centrifuge tube are fractionated; and the radioactivity in each sample is counted with a scintillation counter (merely by adding the sample to a scintillation solvent that accepts alkaline sucrose). All of the radioactivity is found at the meniscus of the centrifuge tube. Taking this to mean that the RNA is replicated discontinuously, the biochemist submits a brief communication of the finding to a scientific journal. The paper is rejected, however, because of an error in the experimental method. What might the error have been?

14-26 A particular operon seems to be regulated by a gene called p. It has been suggested that the p-gene product is not a protein but an RNA molecule that is not translated. The principal support for this idea is that a protein product has never been detected. A series of p mutants is isolated and characterized. Some of these are temperature-sensitive mutants. In an effort to study these thoroughly, the mutations are transduced into strains carrying various suppressors. In two strains, one carrying an amber and one an ochre suppressor, a particular p^- mutation behaves as if it is p^+. You now know whether the p-gene product is a protein or RNA molecule. Which is it and how do you know?

14-27 An operon involved in utilizing a sugar X is regulated by a gene called b. When X is added to the cells, Xase is made; otherwise it is not. If the b gene is deleted (this is denoted by Δb), no Xase can be made. The diploid $b^+/\Delta b$ is inducible. Point mutants of b are of two types; b1 never makes Xase, b2 is constitutive. The partial diploids $b^+/b1$ and $b^+/b2$ are inducible and constitutive, respectively. What is a likely mode of action of the protein encoded by the b gene?

14-28 In a hypothetical *E. coli* operon the regulator gene is closely linked to a region containing two structural genes (consider only one) and an operator. In Table 14-28 the regulator, the operator, and a structural gene are listed in correct sequence. The ability of each of the indicated genotypes to synthesize an enzyme under induced and noninduced conditions is as shown. Which of these

TABLE 14-28

	Phenotype	
Genotype	Inducer absent	Inducer present
$a^- b^+ c^+$	S	S
$a^+ b^+ c^-$	S	S
$a^+ b^- c^-$	s	s
$a^+ b^- c^+/a^- b^+ c^-$	S	S
$a^+ b^+ c^+/a^- b^- c^-$	s	S
$a^+ b^+ c^-/a^- b^- c^+$	s	S
$a^- b^+ c^+/a^+ b^- c^-$	S	S

Note: S = enzyme synthesized in normal quantities; s = little or no synthesis.

genes is the regulator, the operator, and the structural gene? Explain the reasons for your selection.

14-29   A map of a region of the *E. coli* chromosome is shown in Figure 14-29. The arrows denote the direction of transcription. Abbreviations are: *arg*, arginine; *lac*, lactose; *pro*, proline; *tsx*, adsorption site for phage T6; *pur*, purine; *y*, *lac* permease; *z*, b-galactosidase; *o*, an operator; *i*, repressor; *a* and *b*, genes in the purine operon. From a Lac⁺Pro⁺Tsx-s *E. coli* strain, a single-step mutant with the phenotype Lac⁻Pro⁻Tsx-r is isolated. This mutant has the following properties: (1) In the absence of lactose, permease is synthesized if purines are absent, but not if they are present. (Note: The *pur* operon is repressed when purines are present.) (2) If purines are present, permease cannot be induced by lactose. (3) No β–galactosidase activity is observed under any conditions, and this property accounts for the Lac⁻ phenotype. Explain the properties of this mutant.

FIGURE 14-29

arg	*lac*	*pro*	*tsx*	*pur*

y z o p i ← b a o ←

14-30   In *E. coli* the enzyme alkaline phosphatase (APase) is coded for by the gene *phoA*. If phosphate, a product of the reaction catalyzed by APase, is added to a cell culture, synthesis of APase is greatly inhibited in the growing cells. There are two other genes, *phoB* and *phoR*, which also seem to be involved in regulating the expression of *phoA*; mutations in *phoB* lead to reduced synthesis of APase in the presence or absence of phosphate. Mutations in *phoR* lead to constitutive synthesis of APase (see Table 14-30). We know that the *phoB* and *phoR* genes each produce a protein, so the mutations shown in Table 14-30 do not represent *cis*-acting mutations. We also know that the *phoA* gene is being regulated by transcriptional control. Assume that no other proteins or small molecules besides the ones mentioned here are involved in regulating the *phoA* gene.

(a) Which one of the regulatory proteins is interacting directly with phosphate to regulate the expression of APase? Briefly explain why you concluded this.

(b) Which one of the regulatory proteins is interacting directly with a site on the DNA? Is this protein a positive or negative effector? Briefly explain.

(c) Explain how a complex between the PhoB and the PhoR proteins might function in regulating the expression of APase. Include the role of phosphate in your explanation.

TABLE 14-30

	Relative level of APase when phosphate is	
Mutants	Absent	Present
$phoA^+ phoB^+ phoR^+$	100	20
$phoA^- phoB^+ phoR^+$	0	0
$phoA^+ phoB^- phoR^+$	5	1
$phoA^+ phoB^+ phoR^-$	100	100
$phoA^+ phoB^- phoR^-$	5	5

Note: Consider the $phoB^-$ to be a point missense mutation in the $phoB$ gene.

14-31 The regulation of an operon responsible for synthesis of X is dependent on a repressor, a promoter, and an operator. In the presence of X, the system is turned off; an interaction between X and the repressor forms a complex which can bind to the operator.
(a) What kinds of mutations might occur in the repressor? Describe their phenotype (in terms of whether the operon is on or off).
(b) Describe the phenotype of a partial diploid with one wild-type and one mutant gene for each mutant gene.

14-32 How many proteins are bound to the *gal* operon when
(a) Galactose is present and glucose is not.
(b) Both galactose and glucose are present.
(c) Neither sugars are present.

14-33 What is the state of the *gal* operon following addition of lactose to a Lac+ cell? (Remember the formula for lactose and the reaction catalyzed by β-galactosidase).

14-34 Bacteria can use some amino acids as carbon soures. Some examples are glycine, proline, tryptophan, histidine, and serine. Would you expect these operons to be subject to catabolite repression?

14-35   In the arabinose operon, what are the phenotypes of the following partial diploids?
(a) $araA^-araB^+araC^+/araA^+araB^-araC^-$.
(b) $araA^-araB^+araC^-/araA^+araB^-araC^+$.
(c) $araA^+araD^-/araA^-araD^+$.

14-36   An Ara⁺ culture is growing in medium containing both arabinose and glucose. Describe the sequence of events (i.e., transcription and protein synthesis) following depletion of glucose.

● 14-37   What is meant by an attenuator?

14-38   Transcription of the *trp* operon can be initiated by the addition of the tryptophan analogue indole propionic acid (IPA).
(a) Upon induction the enzyme anthranilate synthetase (the TryE protein) is made first and the enzyme tryptophan synthetase (its two subunits are the TrpA and TrpB proteins) is made last. Explain the order.
(b) About how long does it take for an RNA polymerase molecule to transcribe the entire length of the *trp* operon? The polymerization rate is 25 nucleotides/second.
(c) When IPA is washed out and replaced with tryptophan, synthesis of these enzymes ceases in the same sequence. Why?
(d) How does IPA cause derepression?

14-39   How many proteins are bound to the *trp* operon
(a) When tryptophan and glucose are present?
(b) When tryptophan and glucose are absent?
(c) When tryptophan is present and glucose is absent?

14-40   The regulation of the synthesis of an enzyme can be accomplished by controlling transcription or translation. Suppose you are studying the kinetics of appearance of an inducible enzyme after the addition of an inducer. The enzyme can be assayed both by its enzymatic activity and by an immunological test which shows that the protein is present. After induction the enzymatic activity does not appear until several minutes after the appearance of immunological reactivity. Give several possible explanations for this phenomenon.

14-41   The pathway for biosynthesis of biotin is given in Figure 14-41A. Arrows represent the enzymes that catalyze each step and the letters above the arrows represent the genes that code for these enzymes. The five genes *a*, *b*, *c*, *d*, and *f* are clustered on the E.

coli map and form the *bio* operon. A detailed map of the *bio* operon looks is shown in Figure 14-41B. In this map, *oL* and *oR* are the leftward and rightward operators of the *bio* operon, and *p* is a promoter that serves all of the genes.

(a) A polar amber mutation in the *f* gene has been isolated. Predict the derepressed activities of enzymes for the five *bio* genes. What effect would a polar mutation in gene *a* have on genes *b, f, c,* and *d*?

(b) A mutant of *E. coli* has been isolated that never produces the biotin enzymes; that is, they are absent even when the mutant is grown in media containing no biotin. This is known to be a mutation in a gene that codes for a repressor of the *bio* operon. Do you expect this mutant to be dominant or recessive to wild-type? In *trans* or *cis*?

(c) A bacterium related to *E. coli* requires biotin since it is not able to make its own; therefore, biotin must be added to the growth medium for this bacterium to grow. However, this bacterium is known to possess the *bio* genes and it can mutate at very low frequency to gain the ability to grow in the absence of biotin; these mutants produce the biotin enzymes constitutively. Suggest an explanation.

FIGURE 14-41A
$$X \xrightarrow{c} PCA \xrightarrow{f} KAPA \xrightarrow{a} DPA \xrightarrow{d} DTB \xrightarrow{b} \text{Biotin.}$$

FIGURE 14-41B
$$\frac{\quad\quad\quad\quad a \;\; o_L \;\; p \;\; o_R \;\; b \;\; f \;\; c \;\; d}{gal \quad att\lambda \quad\quad\quad\quad bio}$$

14-42   Define stringent and relaxed, when referring to ribosome synthesis.

14-43   (a) What would happen to the rate of rRNA synthesis, if chloramphenicol, an inhibitor of protein synthesis, were added to a growing culture?
(b) What would be the response if a Leu⁻ mutant was transferred to a medium lacking leucine but containing chloramphenicol?

14-44   Consider a branched pathway, shown in Figure 14-44, which is regulated by feedback inhibition.
(a) Indicate the enzymes which are subject to feedback inhibition and for each, identify the inhibitor.

(b) Which steps are likely to be catalyzed by a set of isozymes? How many isozymes are probably in these steps?

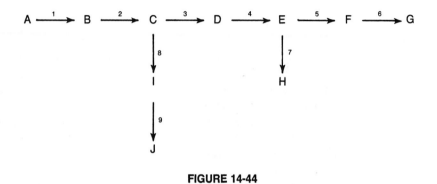

FIGURE 14-44

15

Lytic Phages

15-1 Bacteria are allowed to grow on an agar surface until a confluent turbid layer appears. Then, 10^3 T4 phage are spread on the surface. Six hours later (a time sufficient for plaque formation, if the phage had been added at the time the bacteria were placed on the agar) no plaques are evident. Explain.

15-2 Some types of phages produce small plaques and some produce large plaques.
(a) Give several reasons why this might be so, and explain why populations of smaller phages tend to produce larger plaques.
(b) Even for a given type of phage there will be varying plaque size if phage particles are added to agar containing bacteria. However, if phage and bacteria are incubated together for several minutes before adding agar, the plaques are more uniform in size. Propose an explanation.
(c) Some phage species produce clear plaques with large turbid halos. These halos tend to increase in size even after phage production has stopped. Furthermore, very few phage are recovered from the halo region. Suggest a cause for the halo.

15-3   If a culture of *E. coli* is infected with T4, the burst size is about 300, whereas with T7 it is about 100. For each, the lysis time is about 22 minutes at 37°C. However, T7 makes plaques whose diameter is roughly four times as great as T4 plaques. Propose an explanation.

15-4   Typically, a bacterium infected with a virulent phage produces from 50 to 500 phage per cell. If a phage mutant could produce only one new phage particle per cell, clearly a plaque could not be produced. Normally 10^8 bacteria are put on an agar surface when assaying for phage by the plaque method. The maximum bacterial concentration reached is about 10^{10} bacteria per petri dish. Assuming that the doubling time of the bacterium and the length of the phage life cycle are the same, what is the minimum burst size that a phage mutant can make and still produce a visible plaque? The smallest visible plaque contains about 10^5 phage.

15-5   The number of phage in a lysate is being determined by plating. First, 0.1 ml of the lysate is diluted in 10 ml of buffer (tube A). Two successive dilutions of 0.1 ml to 10 ml are made (tubes B and C), followed by two successive dilutions of 1 ml to 9 ml (tubes D and E). Then, 0.1 ml from various dilution tubes is placed in agar with sensitive bacteria, and the next day, plaques are counted. The number of plaques obtained from the 0.1 ml aliquots from tubes C, D, and E are 2010, 352, and 18, respectively. What is the best estimate of the number of phage per ml in the lysate? Why do the number of plaques per dilution tube not simply follow the dilution?

15-6   Phage T4 normally forms small clear plaques on a lawn of *E. coli* strain B. A mutant of *E. coli* called B/4 is unable to adsorb T4 phage particles so that no plaques are formed. T4*h* is a host-range mutant phage capable of adsorbing to *E. coli* B and to B/4 and forms normal looking plaques. If *E. coli* B and the mutant B/4 are mixed in equal proportions and used to generate a lawn, what will be the appearance of plaques made by T4 and T4*h*?

15-7   In a broth containing glucose and yeast extract, *E. coli* grows with a doubling time of 30 minutes. Coliphage T7 has a latent period of 20 minutes under these conditions and a burst size of 200 phage per infected cell. If a culture of 2 x 10^7 *E. coli* per ml is growing exponentially and 5000 plaque-forming units per ml of T7 phage are added, when will the culture lyse? A few minutes before visible lysis a certain number of bacteria will remain uninfected. The estimate for uninfected cells a few minutes before the time of lysis would be difficult to confirm experimentally, since plating of cells and phage together might lead to infection and killing of all the surviving bacteria. In practice, about 100 colony-forming units per ml are found. Some of the colonies are large and round. What might these colonies represent? (In working with this problem assume that phage adsorption is instantaneous and that multiply-infected bacteria give the same burst as singly infected bacteria.)

15-8   In an experiment in which 10^7 bacteria are mixed with 5×10^7 phage, how does one determine the actual multiplicity of infection and the number of infective centers?

15-9   One ml of a bacterial culture at 5×10^8 cells/ml is infected with 10^9 phage. After sufficient time for greater than 99 percent adsorption, phage antiserum is added to inactivate all unadsorbed phage. The infected cell is mixed with indicator cells in soft agar, and plaques are allowed to form. If 200 cells are put in each petri dish, how many plaques will be found?

15-10   Consider a phage with the strange property that productive infection occurs only if at least two phage adsorb to the bacterium. If you have 10^8 bacteria and add to this culture 3×10^8 phage, how many bacteria will be productively infected?

15-11   Suppose a culture is to be simultaneously infected with phages A and B. What ratio of phage to bacteria must be used to ensure that 90 percent of the bacteria are infected with at least one phage of each type?

15-12   An *E. coli* culture is infected with T2 and T2*h* (host-range mutant) at a ratio of 5 phage of each type per bacterium. A lysate is obtained, a portion of which is adsorbed either to (1) wild-type *E. coli* or (2) *E. coli* B/2 (unable to adsorb T2)—in each case, at a multiplicity of infection of 0.1. Antiserum is added to remove unadsorbed phage and then the infected cells are plated on a lawn of B/2. Twice as many plaques are formed with infection (1) than with infection (2). Explain the difference.

15-13   A lysate of phage R3 is found to have the following peculiar property. By plating on sensitive bacteria it is found that there are 10^8 plaque-forming units per ml. If 0.1 ml of the lysate is mixed with 10^8 bacteria, there are 10^7 infective centers. However, if the bacteria are infected at moi = 3 with a mutant p^- and with 0.1 ml of the lysate described above, 3×10^7 infectious centers result. If the bacteria are infected at moi = 3 with a mutant q^- with 0.1 ml of the lysate, 10^7 infective centers are found. Explain the difference in the number of infective centers.

15-14   P2 and P4 are bacteriophages of *E. coli*. They have the following properties: (1) When one P2 phage infects a bacterium, the bacterium usually bursts, giving about 100 P2 progeny. (2) When a P4 phage infects a bacterium, the bacterium survives because P4 is a defective phage. (3) When P2 phage and P4 phage coinfect the same bacterium, lysis of the bacterium gives 100 P4 progeny and no P2 progeny (because P4 inhibits the growth of P2). If 3×10^8 P2 and 2×10^8 P4 are added to 10^8 bacteria, then:
(a) How many bacteria will not be infected at all?
(b) How many bacteria will survive?
(c) How many bacteria will produce P2 progeny?
(d) How many bacteria will produce P4 progeny?

•15-15   List the main stages of a phage life cycle.

•15-16   Name the minimal (essential) structural components of the simplest phage particle.

•15-17   Give several examples of mechanisms by which a phage converts a bacterium to a phage-producing machine.

15-18   What would you expect to happen if, after virulent phage infection (such as with T4), all the phage genes were transcribed and translated at once following phage DNA infection?

15-19   Rank the *E. coli* T phages in order of increasing molecular weight of their DNA.

•15-20   What kinds of nucleic acids are found in phage particles of difference phage species?

15-21   What is meant by a nonessential gene? Can a phage gene be truly nonessential?

15-22   What is meant by term phage–host specificity and what is the most frequent cause of this specificity?

15-23   A new protein X appears in infected cells. Describe various experiments that you might perform to prove that the gene coding for X is phage-encoded and is not encoded in host DNA.

15-24   How would you show whether or not a phage-encoded gene product is required throughout the infectious cycle or only at a unique time?

●15-25   What is meant by Sup$^+$ and Sup$^-$ hosts with respect to phage mutant growth?

15-26   Name four features by which T4 DNA and T7 DNA differ.

15-27   In an infection with T4 phage, which of the following are true (several answers are):
(a) *E. coli* DNA is degraded by a T4-encoded enzyme.
(b) T4 phage DNA is replicated by *E. coli* polymerase III.
(c) Deoxycytidine triphosphate (dCTP) is converted by cytidine hydroxymethylase to dHTP, which is then glucosylated by α-glucosyl transferase, and the glucosylated dHTP is then incorporated into T4 DNA by a DNA polymerase.
(d) Early mRNA is made by using *E. coli* RNA polymerase with an *E. coli* σ factor.
(e) Late mRNA synthesis is delayed until the T4 DNA has replicated.

15-28   If an *E. coli* culture is simultaneously infected by phages T4 and T7, each at a multiplicity of infection of 5, only T4 phage will be produced. From what you know about T4 biology, propose a simple explanation. What would you expect to happen if the T4 was added 10 minutes after addition of T7?

15-29   What is the function of the T4 dCTPase (deoxycytidine triphosphatase)?

15-30   Describe the course of T4 phage DNA synthesis following infection of *E. coli* with a T4 mutant which cannot synthesize (a) cytidine hydroxymethylase, or (b) α-glycosyl-transferase.

15-31   Can one determine whether the early genes or the late genes of phage T4 are injected first into a host bacterium? Explain your answer.

15-32 List some T4 gene products that would be expected to act catalytically; list those products that would be required in stoichiometric amounts.

15-33 *E. coli* phage T4 synthesizes many enzymes—for example, thymidylate synthetase and deoxycytidine deaminase—which are not essential for growth. Suggest how such genes may have evolved as part of the T4 genome.

15-34 What is the basic principle used by T4 in regulating the transcription sequence of the mRNA molecules made before DNA replication begins?

15-35 If a T4 phage has a large deletion which of the following will be true?
(1) The activity of one or more proteins will be greatly altered or altogether missing.
(2) The phage DNA will be smaller.
(3) The terminal redundancy will be larger.
(4) The phage DNA will be the same size.
(5) Cyclic permutation will be eliminated.

15-36 Suppose you had a phage whose linear DNA is synthesized by the rolling circle mode and is packaged by "the headful rule" (that is, DNA is added to a head of fixed size until no more DNA can fit). However, the DNA is normally neither terminally redundant nor cyclically permuted. You find a mutant strain of this phage, the DNA of which has a deletion in a nonessential gene. This phage is used to infect a bacterium, and many phage are produced. The DNA is isolated and is treated with an exonuclease, which removes a few bases from the 5'-P end. The treated DNA is then exposed to conditions which could circularize T4 DNA (if it were also pretreated with the exonuclease). This DNA is examined by the electron microscope. Would circles be found?

15-37 Most double-stranded DNA phage have several classes of mRNA that can be divided into two major groups—early mRNA and late mRNA. The genes carried on the early mRNA species vary from one phage to the next. Nonetheless, there are certain genes that are usually on early transcripts and some which are invariably on late transcripts. What are these genes?

15-38 Phages T4, T7 and λ use different modes of regulating transcription. What are they?

15-39 The Meselson-Stahl experiment showed that in DNA replication in
       E. coli, each DNA molecule produces two daughter molecules, so
       that the total number of bacterial DNA molecules in a growing
       culture increases exponentially. Given the fact that the rate of
       nucleotide addition during T7 DNA replication is the same as that
       for *E. coli* DNA (in an uninfected cell), is it possible that T7
       DNA replicates by successive doubling? If so, what would have to
       be true of the timing of initiation? Some useful numbers are the
       molecular weight of *E. coli* and T7 DNA, the length of the life
       cycle of T7, the normal burst size (about 200), and the time
       required to replicate *E. coli* DNA (40 minutes).

15-40 T4, T7 and λ all have life cycles in which a large fraction is
       devoted to late transcription. Why is the duration of time
       allotted to late transcription greater than the time for early
       transcription?

15-41 Bacteriophage T5 particles labeled with radioactive DNA are
       adsorbed to *E. coli* in the presence of cyanide, which stops the
       generation of a cellular energy supply. After adsorption at 37°C
       for 20 minutes in the presence of cyanide, the infected cells are
       agitated in a blendor to remove the phage. The infected cells now
       contain radioactive DNA, which can be extracted and which
       sediments as though its molecular weight were 10 percent that of
       a T5 DNA molecule. In a second experiment, wild-type T5 is
       adsorbed to Sup⁻ cells in the presence of cyanide and then
       removed by blending. The infected cells are superinfected
       separately with various T5 amber mutants. Progeny phage are
       produced only with mutants in two of the twenty known T5 genes.
       These two genes are both at one end of the genetic map. From
       these data what can you conclude about the mode of injection of
       T5 DNA and the structure of T5 DNA?

15-42 Although most DNA molecules contained in phage heads are line-
       ar (PM-2, and ϕX174 are exceptions), circular replicating forms
       are common. Describe several methods that might be used to
       generate such circles.

15-43 What enzymes are needed to form a supercoiled λ DNA molecule
       after injection? State them in order of activity.

15-44 Shortly after injection of λ DNA and prior to any transcription,
       dimers of λ DNA are found. The frequency increases with the

multiplicity of infection. This occurs in cells lacking all known recombination functions. How do these dimers probably form?

15-45   Which transcripts of phage λ are made by an N^- mutant?

15-46   What functions are lacking in a λQ^- mutant?

15-47   A deletion of λ called *nin5* removes a large segment of DNA between genes *P* and *Q*. This deletion allows growth of a λ phage that is mutant in an important early gene. Which gene is it?

15-48   If an *E. coli* culture is heavily irradiated with ultraviolet light and then infected with T4 phage, the burst size of T4 is nearly normal. If it is instead infected with λ phage, very few phage are produced. Explain this difference.

15-49   Revertants of some λO^-(Am) mutants have been obtained by plating these mutants on a Sup$^-$ host and picking the few plaques that occur. These revertants frequently carry a mutations in gene *P*. Revertants of some λP^-(Am) mutants having mutations in gene *O* have also been isolated. Furthermore, λ cannot plate on an *E. coli dnaB* mutant but phage mutants have been isolated that can grow; these have a mutation in gene *P* but never in gene *O*. What can you conclude from these genetic results?

15-50   Suppose a λ phage has infected *E. coli* and is actively replicating its DNA. No DNA has been packaged, since late mRNA has not yet been made. If a T4 phage is then added, will the λ phage still be made?

15-51   There exist certain transducing λ particles that carry the genes for synthesizing tryptophan. The locus of the *trp*-gene insertion is shown in Figure 15-51. The tryptophan (*trp*) operon is transcribed in the direction indicated by the horizontal arrow—that is, in the same direction as the phage mRNA that goes leftward from *pL*. There are several classes of tryptophan-transducing particles that differ according to the size of the DNA molecule replaced by *E. coli* DNA. In each class, the size of the inserted bacterial DNA is the same size as the DNA which is absent, the bacterial DNA contains the tryptophan synthetase gene, the insertion begins at *att* and moves to the right, and no essential phage genes are missing. These particles have the property that when they infect a bacterium lacking tryptophan synthetase, this enzyme is made.

(a) If the phage carries a point mutation in the λ *N* gene, no tryptophan synthetase is made. Why not? What part of the *trp* operon must be missing for this to be the case?
(b) Consider a transducing particle whose piece of bacterial DNA is so large that part of the *N* gene is replaced by bacterial DNA. Will any tryptophan enzymes be made? Explain briefly.

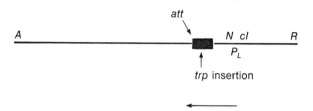

FIGURE 15-51

15-52   The λ *ti⁻* mutant makes small plaques and has a burst size about 10 percent that of wild-type λ. In a mixed infection of *ti⁻* and *ti⁺* no *ti⁻* phage are produced. By use of a density label it is known that in the mixed infection *ti⁻* has failed to replicate. Furthermore, if the *ti⁺* phage is also O^-P^- (that is, mutant in the only phage genes needed for λ DNA replication), in a mixed infection with $ti^-O^+P^+$ and $ti^+O^-P^-$ again no *ti⁻* phage appears in the burst; the yield of $ti^+O^-P^-$ is that of the wild-type. What kind of defect is the *ti⁻* mutation?

15-53   How many λ particles can be packaged from a dimer circle? A trimer circle?

15-54   If λ*P⁻* infects *E. coli* at a low multiplicity of infection, no phage are produced because there is no DNA replication. If the multiplicity of infection is ten and either the bacterial or phage recombination systems are active, about one-third of the infected cells release one phage. Explain.

15-55   The genes *J* and *E* of phage λ encode a tail protein and the major head protein respectively. The *E⁻* and *J⁻* mutations are amber mutations and the bacteria used in the following experiments do not contain a suppressor. If *E. coli* is infected with an *E⁻* phage, no viable phage are produced but lysis occurs. Such a lysate is called an E⁻ lysate. A similar infection with λ *J⁻* yields a J⁻ lysate, which also contains no viable phage. If E-

and J- lysates are mixed, viable phage form in numbers reaching about 50 phage per original cell. To study this phenomenon the following procedure is carried out. The E^- lysate is mixed with *E. coli* which has an amber suppressor, and the bacteria are then removed by centrifugation. The supernatant is devoid of activity—that is, complementation with the J- lysate is abolished. The sediment, however, which contains all of the added cells, is active, in the sense that if the J^- lysate is added, the cells develop phage and lyse.

(a) What is happening in this experiment?

(b) What is the genotype of the phage produced by the infective centers?

15-56   *E. coli groE⁻* mutants do not allow phage to form active head particles. Recently a special λ phage variant has been isolated which can grow on *E. coli groE⁻* host strains. This phage was constructed by recombinant DNA techniques starting with a phage strain (called phage A) whose DNA can be cleaved by the EcoRI nuclease only at the three sites shown in Figure 15-56A by

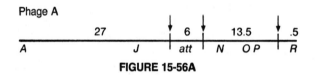

FIGURE 15-56A

vertical arrows; the distances separating the sites are given in kilobase pairs and some λ genes are indicated. If the cleaved fragments are annealed at 4°C and then treated with DNA ligase, some intact λ DNA molecules form which can transfect appropriately treated bacteria and produce phage. The fragments can also join with other fragments of DNA (from *E. coli* for instance) and if the unit formed is the right size and if the two terminal fragments are present, again a functioning phage DNA molecule can result; that is, the central fragment, which contains nonessential genes can be replaced by a piece of "foreign" DNA, as long as the piece is not too big. λ DNA fragments were incubated either alone or in the presence of EcoRI nuclease fragments from *E. coli gro⁺* (wild-type) or *E. coli groE⁻* DNA. The annealed mixtures were treated with DNA ligase, and then used to transfect *E. coli gro⁺* as well as *E. coli groE⁻*. The following three mixtures give rise to plaque-forming phage, as measured on *E. coli gro⁺*: (1) λ DNA fragments alone, (2) λ DNA

fragments + *E. coli gro⁺* DNA fragments, and (3) λ DNA fragments
+ *E. coli groE⁻* DNA fragments. Only mixture (2) gives rise to
plaques which can plate on *E. coli groE⁻*; one of these plaques
(called phage B) is picked and the phage in it are grown to a
high titer stock. The DNA is extracted, denatured, and reannealed
with denatured DNA from phage A. Figure 15-56B shows the
heteroduplex structure that is commonly observed. After
mutagenesis of phage B, it is possible to isolate some mutants
which form plaques on the following bacteria: *gro⁺*, *gro⁺supF*, and
groE⁻supF, but not *groE⁻*, where *supF* is an amber suppressor. One
of these mutants was selected for study and called phage C.
Phages A, B, and C were each used to infect *E. coli uvrA⁻*, which
had been preirradiated with ultraviolet light in order to abolish
the capacity for host protein synthesis. Radioactive amino acids
were administered during infection in order to label the phage
proteins synthesized, and these proteins were separated by
electrophoresis in acrylamide gels containing 1 percent sodium
dodecyl sulphate, and phage protein bands were detected by
autoradiography. Phage B cause synthesis of a 65,000 dalton
protein (protein X), which is not made by either phage A or phage
C. Phages B and C synthesize a protein of 15,000 daltons (protein
Y), which is not made by phage A. Phage A synthesizes a protein
of 20,000 daltons (protein Z), which is not made by phages B
and C.

(a) Is the *groE* mutation dominant or recessive to the wild-type
allele? Give pertinent evidence.
(b) What is the genotype of phage B?
(c) What is the genotype of phage C?
(d) What is protein X?
(e) What is protein Y?
(f) What is protein Z?

FIGURE 15-56B

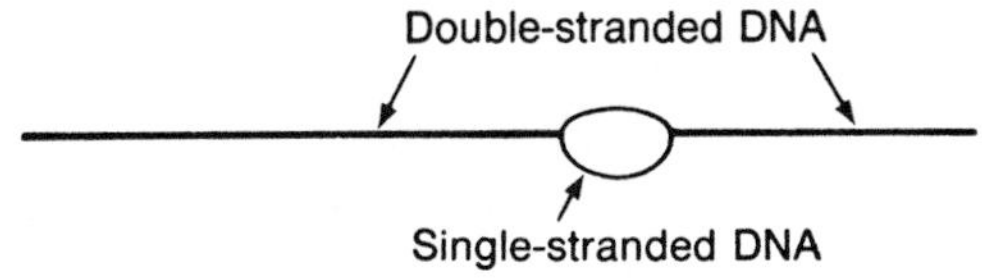

Double-stranded DNA

Single-stranded DNA

15-57   (a) For the single-stranded DNA phages what is meant by the
         (+) strand and by the (−) strand?
         (b) Is the meaning the same for the RNA phages?
         (c) What is the difference with respect to transcription of the
         (+) strands of the DNA and RNA phages?

15-58   (a) For *E. coli* phage ∅X174, would you expect the (+) and (-) circular, single-stranded DNA strands of the phage (isolated from the double-stranded replicative form) each to be infective in transfection experiments?

(b) Would you also expect the (+) and (-) RNA strands of the *E. coli* RNA phages to be individually infective?

15-59   (a) How does ∅X174 manage to include all of its necessary genetic information in such a small DNA molecule?

(b) Do you think that ∅X174 could reproduce in a heavily UV-irradiated host?

15-60   Which phage would be expected to produce clearer plaques, ∅X174 or M13? Why?

15-61   When phage DNA enters a bacterium, it can usually be reisolated in the laboratory as free DNA or occasionally bound to what is thought to be a membrane fraction. Where would you expect to find the RNA of f2, R17 or MS-2?

16

Lysogenic Phages and Transduction

• 16-1   Define temperate phage, lysogen, defective lysogen, integration, excision.

16-2   The *cI* gene of λ codes for the immunity repressor. If the λ is *cI*⁻, the plaque is uniformly transparent. If it is *ci*⁺, the plaque has a turbid center. Explain the plaque morphology. Would you expect T2 plaques ever to have turbid centers?

• 16-3   Two temperate phages are homoimmune. What phage elements must they have in common?

16-4   Suppose you have a λ mutant that makes a clear plaque on *E. coli* strain A and a turbid plaque on strain B. How can this difference be explained?

16-5   An *int*⁻ mutant can integrate if the cell is coinfected with an *int*⁺ phage, the latter "helper" phage provides the missing function.
(a) Could a phage with an operator mutation (e.g., λvir) be helped to integrate?

(b) Could a cII^- or $cIII^-$ mutant be helped to integrate?

(c) Is there any circumstance in which a lysogen could contain a cI^- mutant? (This is a little tricky. Do not consider ordinary lysogens).

16-6   Is it possible to form a dilysogen by simultaneously infecting a bacterium with a pair of heteroimmune phages?

•16-7   A normal λ phage cannot successfully infect a bacterium lysogenic for normal λ (that is, it cannot make more phage and lyse the bacterium).

(a) Why not?

(b) Suppose the phage added to the lysogen is a mutant in that the left (oL) and the right (oR) operators are mutated (that is, they are defective in the same way that a *lac* operon o^c mutant is defective). Can this phage successfully infect a wild-type lysogen? Explain briefly.

16-8   If one λ phage particle infects a cell lysogenic for λ, which of the following will probably happen?

(a) A normal phage cycle producing about 50 phage particles will occur.

(b) The λ DNA will circularize but will not replicate.

(c) The cell will die.

(d) The λ prophage will be excised.

(e) The λ DNA will not be injected.

16-9   If a λ lysogen is induced with ultraviolet light, a lysate results in which about one clear-plaque mutant per 10^4 phage is present. If a culture of a lysogen is grown without any prior induction and the culture fluid is analyzed for free phage that are spontaneously released, the fraction that form clear plaques is always higher than 10^{-4}.

(a) Explain the higher frequency of clear-plaque mutants among the spontaneously released phage.

(b) Among the spontaneously released phage in which of the genes *cI*, *cII*, or *cIII* do you suppose the mutations reside that yield a clear plaque?

16-10   A temperate phage has the gene order $A\ B\ C\ D\ E\ F\ G\ H$ and a prophage gene order $G\ H\ A\ B\ C\ D\ E\ F$. What information does this give you about the phage?

16-11 $F'gal(\lambda)bio$ circular, double-stranded DNA is nicked randomly with pancreative DNase at an average frequency of one nick per double-stranded molecule. The DNA is denatured and the circular single strands are isolated. One single strand of circular $F'gal(\lambda)bio$ DNA is renatured with the complementary single strand of λ phage DNA. What structures may be formed?

16-12 A λ phage genetic map (as obtained in standard crosses) containing only a few of the known genes is approximated below: Which of these genes would show the highest frequency of cotransduction with the *gal* gene if a P1 phage grown in a *gal*+ λ-lysogenic cell were used to transduce a *gal*-(λ) lysogen?

A   J   cl   P   Q   R

FIGURE 16-12

16-13 You have isolated two independent mutant strains of the phage λ; both of these mutants form clear plaques. When either mutant alone infects *E.coli* at moi = 10, no lysogenic survivors can be isolated. However, when *E. coli* is coinfected with 5 phage per cell of the first mutant plus 5 phage per cell of the second mutant, then 10 percent of the cells survive the infection, and almost all of these survivors are lysogenic. When the lysogenic survivors are induced, 99 percent of the plaques produced are clear.

(a) What λ function or functions do you think are eliminated in these mutants?

(b) What is the most likely explanation for the increased frequency of lysogenization which occurs during coinfection by the two mutants?

16-14 Which of the following λ mutants cannot lysogenize at normal frequency? N^-, P^-, Q^-.

16-15 Phage variants have been prepared containing many different attachment sites.

(a) What attachment sites are in λ *bio* and λ *gal* phage?

(b) What attachment sites are generated by Int-mediated recombination between λ *bio* and λ *gal*?

16-16 Which of the following pairs of phages can undergo Int-promoted recombination as shown? λ *bio* x λ *bio*; λ *gal* x λ *gal*; λ+ x λ *bio*; λ *gal* x λ+.

16-17   Does an *int⁻* mutant of λ make a clear or a turbid plaque?

16-18   A λ mutant is isolated which is defective in that, though it forms plaques normally, a lysogen containing a single mutant prophage fails to produce phage when induced. The mutant lysogenizes with normal frequency. In studying this mutant, you infect bacterial strain R with this mutant at a moi = 3. Strain R has been obtained by lysogenizing a Gal⁻ bacterium containing a segment of a λ prophage that has only the right-pro-phage attachment site, with a *gal⁺* transducing phage. The resulting structure of this bacterium is shown in Figure 16-18, in which the light line represents the λ prophage. Strain R contains a normal cI repressor and therefore is immune to superinfection. When R is infected with the λ mutant and the infected cells are plated on an indicator agar (galactose tetrazolium) in which Gal+ and Gal- colonies are white and red, respectively, roughly equal numbers of white and red colonies are found. In what gene does the λ mutant map and what type of mutant is it?

| *gal⁻* | *PP'* | | *gal⁺* | *BB'* | *bio* |

FIGURE 16-18

16-19   (a) Phage λ infects *E. coli* at a total multiplicity of 10. An extract is made from these infected cells and the DNA-binding activity is measured in an *in vitro* system. DNA-binding activity is a measure of the concentration of the cI repressor in the infected cells. The data in Table 16-19 are obtained. Why is the

TABLE 16-19A

Infecting phage	*DNA binding activity*
cI^+	4.0
cI^-	0.1
$cI^- + cIII^-$	2.0
$cI^- + cII^-$	2.0
$cI^- + y^-$	0.1
$cI^+ + cI^- cII^-$	2.0
$cI^+ + cI^- y^-$	2.0

DNA-binding activity low in two of the infections? How can the DNA of λ*imm434* be used in these experiments?

(b) The Cro protein shuts off the cI repressor. If a lysogen containing λ*N⁻O⁻cI857* is grown at 42°C, thus inactivating the temperature-sensitive *cI857* repressor, the Cro protein is made constitutively. Explain the data given in Table 16-19B which shows the frequency of lysogeny of the immλ phage when a *cro*-constitutive lysogen is infected by the indicated phage mixtures at a multiplicity of infection of five each.

TABLE 16-19B

Infecting phage	*Percent lysogeny (immλ phage)*
immλ cI⁺	1
immλ cI⁺ + imm434 cII⁻	1
immλ cI⁺ + imm434 cIII⁻	1
immλ cI⁺ + imm434 cI⁺	45

16-20   When Hfr males conjugate with F^- cells lysogenic for λ, zygotes normally survive. However, when Hfr males lysogenic for λ conjugate with F^- nonlysogens, zygotes produced from matings that have lasted for almost two hours lyse, owing to the zygotic induction of λ.

(a) How can you explain zygotic induction?

(b) How can you determine the locus of the integrated λ prophage?

16-21   Remember the *E. coli* gene order *gal att bio*, in which *att* = *BOB'* is the attachment site in a nonlysogenic bacterium. If phage P1 is grown on a nonlysogen, some of the transducing particles carry the *gal* gene, some carry *bio*, and some carry *gal* and *bio* genes. The ratio of particles transducing both *gal* and *bio* together to those transducing *gal* is 0.01. If the P1 phage is grown instead on a λ lysogen, would you expect the number of particles transducing the *gal* and *bio* loci together to increase, decrease, or stay the same? Why?

16-22   A λ phage mutant called λ*b2* cannot lysogenize because it has a deletion starting just to the left of the *O* region of the attachment site—that is, its attachment site is $\Delta OP'$, in which Δ represents a deletion. In a mixed infection with λbio using conditions favoring the lysogenic response, dilysogens containing both $\lambda b2$ and λbio are found. Very few single lysogens (these can be ignored in your thinking) and no *(b2,b2)* or *(bio,bio)* dilysogens are found. How have the *(b2, bio)* dilysogens formed?

16-23   Replication of λ DNA can be detected by hybridization of labeled λ mRNA with all of the DNA isolated from an infected cell or from an induced lysogen. As replication proceeds and the number of copies of λ DNA increases, more λ mRNA can be hybridized. Interpret the following observations.
(a) If an *xis*⁻ lysogen is induced, the amount of hybridizable λ mRNA increases with time.
(b) Also, using *bio* mRNA and *gal* mRNA (both of which are easily obtainable) the amount of these mRNA's that can hybridize also increases with time.
(c) If an *int*⁺ lysogen is induced, the amount of hybridizable *bio* and *gal* mRNA (as well as λ mRNA) increases, though less of these hybridize than in part (b).

16-24   Infection of a λ lysogen with ten λ phage particles does not result in phage development. This is the immune response. If the multiplicity of infection is 30, phage development occurs. In a dilysogen 60 phage particles per cell are needed to initiate a successful infection. Explain.

16-25   The temperature-sensitive λ cI mutant *cI857* is especially useful, because a lysogen need only be heated to 40°C to induce phage development, and induction is a convenient way to grow λ. The *cI857* repressor is reversibly denaturable in that if, after heating, a cell is returned to 34°C or less, the repressor renatures and becomes active.
(a) At 40°C is the *cI* gene of a lysogen still transcribed?
(b) If the temperature is returned to 34°C, will the *cI* gene be transcribed?
(c) If a growing λ *cI857* lysogen is heated for 10 minutes and then cooled, phage development occurs on continued incubation in growth medium. If the lysogen is instead cooled after 5 minutes, there is no phage development and the bacteria survive. Explain.
(d) In (c) when there is no phage development, if the bacteria are plated to form colonies, the colonies are found to consist of nonlysogenic cells. Explain.

16-26   Suppose you have a tandem double lysogen (that is, two prophages are next to one another) and both prophages are *int⁻* and are incapable of DNA replication. Also, both prophages contain a mutant repressor that is temperature-sensitive, so that at high temperature mRNA can be made. If this double lysogen is heated to the high temperature, which of the following will occur? (1) No phage will be produced; (2) one phage will be produced per cell; (3) two phage will be produced per cell; (4) only transducing phage will be produced.

16-27   You have given a friend a mutant λ phage that is defective in gene *J* and in one of the *red* genes needed for phage-mediated generalized recombination. She complains that growth of the *J⁻red⁻* phage in *rec⁻* cells lysogenic for a *P⁻red⁻* prophage yields *J⁺P⁺* recombinants (the *red⁻* mutations carried by the two phage types are identical). Can you explain this "extraordinary" recombination? How might you show by a genetic experiment that your explanation is probably correct? Hint: Draw a picture of the integration reaction with the superinfecting *J⁻red⁻* DNA (assuming *POP'* x *POB'* and *POP'* x *BOP'* are allowed reactions) and consider that the normal λ maturation process applies to this situation.

16-28   In an *E. coli* tandem dilysogen, the left and right prophages have the genotypes *cI857int⁻P⁻* and *cI857int⁻A⁻*, respectively. The repressor of each is inactive at 42°C and the prophages lack a functional integrase. The mutations *P⁻* and *A⁻* are genotypic markers whose locations are shown on the λ genetic map (Appendix C). If the lysogen is heated to 42°C, what will be the genotype or genotypes of the phage progeny? Why will more phage particles be produced than in the experiment described in problem 16-26?

16-29   (a) The λ amber mutant *A32* fails to plate on *E. coli* strain 594 since this strain is *sup⁻*. The λ mutant *int6* is defective in normal prophage excision. The λ *cI857* mutation renders the cI repressor inactive at 42°C, whereas it has normal activity at 32°C. You have just lysogenized strain 594 with λ*cI857int6A32* and have a collection of lysogens. Design a genetic test to distinguish the single lysogens from the dilysogens. It is known that 40 percent of all lysogens are tandem dilysogens and 60 percent are single lysogens.
(b) On another occasion, *E. coli* strain B583 is lysogenized. This strain contains a sex plasmid on which the λ attachment site is

located. This site is deleted from the chromosome, so a prophage can be inserted only in the plasmid. What simpler genetic test could now be used to distinguish the single lysogens from the dilysogens?

16-30 A temperate phage is treated with the mutagen nitrosoguanidine and five clear-plaque mutants are picked for further study. To study the lysogenic pathway of phage development, the clear mutants are coinfected pairwise with each other and with the turbid wild-type phage c^+. The frequency of lysogeny is measured, with the results shown in Table 16-30.
(a) How many cistrons are there? Which mutations are in the same cistrons?
(b) Phage mutant $c5$ is known never to lysogenize its host. What is a possible function of the wild-type product of the gene defined by the $c5$ mutation? For example, is its function to establish or to maintain lysogeny?
(c) Why do the mutants $c1$ through $c5$ form clear plaques, whereas c^+ forms a turbid plaque?
(d) If a lysogen of this wild-type phage is infected with an un-related temperate phage, what will be the outcome?

TABLE 16-30

	Percent lysogeny					
	$c1$	$c2$	$c3$	$c4$	$c5$	c^+
$c1$	2	59	57	1	50	52
$c2$		1	0	56	52	60
$c3$			1	57	58	55
$c4$				1	56	57
$c5$					0	55
$c6$						60

16-31 Mutants of a phage K are known which affect the immunity repressor of the phage and are therefore totally unable to lysogenize. These mutants have been mapped in the tul gene. In addition, mutants of K can be isolated which lysogenize rarely. These mutants map in the $dilA$ and $dilB$ genes. Furthermore, it is known from other experiments, that dil mutants are not defective in the enzymes which catalyze the insertion of K into the chromosome. The data for frequency of lysogeny for each of these mutants are shown in Table 16-31. Based on the foregoing data, answer the following:

(a) What process is going on when mixed infections of tul^- and $dilA^-$ or $dilB^-$ yield lysogens?

(b) Provide a reasonable model for the role of the dil genes in lysogenization.

(c) Why are single lysogens of tul^- not isolated after mixed infection, while single $dilA^-$ or $dilB^-$ lysogens can be found?

TABLE 16-31

Infecting phage or phages	Relative percent of lysogenic survivors	Genotype of prophage
$tul^+ dilA^+ dilB^+$	100	$tul^+ dilA^+ dilB^+$
tul^-	0	—
$dilA^-$	0.1	$dilA^-$
$dilB^-$	1.0	$dilB^-$
$tul^- + dilA^-$	60	$dilA^-$ or $tul^- dilA^-$ dilysogen
$tul^- + dilB^-$	75	$dilB^-$ or $tul^- dilB^-$ dilysogen
$dilA^- + dilB^-$	80	$dilA^-$ or $dilB^-$ or $dilA^- dilB^-$ dilysogen

16-32 Two investigators meet each other at a scientific meeting and discover that they have been studying two similar phages. Both phages are temperate and have the same appearance in the electron microscope. The phages differ in their host range (ability to adsorb to different *E. coli* strains); each carries a distinct immunity (repression) system; and each attaches to different sites on the *E. coli* chromosome. The properties of these two phages are compared in Table 16-32.

(a) The investigators want to create a hybrid phage carrying the immunity specificity of phage 186 and the host range (adsorption properties) of phage P2. Design a selection for such a phage.

(b) Having obtained $hP2\ imm186$ (the desired hybrid), the investigators wish to determine its chromosomal attachment site. How can this be done by bacterial mating?

(c) How can the chromosomal attachment site be located by transduction?

(d) In the experiment of part (c) 95 percent of the hybrid phage $hP2\ imm186$ tested attach near the *phe* gene. What does this mean?

TABLE 16-32

Property	Phage	
	186	*P2*
Ability to adsorb to *E. coli* K	+	+
Ability to adsorb to *E. coli* C	−	+
Ability to form a plaque on *E. coli* K(186)	−	+
Ability to form a plaque on *E. coli* K(P2)	+	−
Location of prophage attachment site	Near *phe*	Near *ile*

16-33 Suppose you are studying a newly isolated temperate phage. The phage can integrate into the bacterium DNA very near the *lac* genes. You isolate nonsense mutations and classify them into four complementation groups *a*, *b*, *c*, and *d*.

(a) After lytic growth of various pairs of mutant phages you measure the recombination frequency and obtain the data shown in Table 16-33A. From these data, draw the recombination map of the vegetative (lytic) phage.

(b) For the prophage, you measure P1 cotransduction of *lac+* and the wild-type alleles of bacteria having each of the following genotypes: *a−*, *b−*, *c−*, and *d−*; for example, P1 grown on a strain whose genotype is $lac^+a^+b^+c^+d^+$ is used to transduce a strain whose genotype is $lac^-a^-b^+c^+d^+$ and the frequency of appearance of transductants having the genotype $lac^+a^+b^+c^+d^+$ is measured. The results are shown in Table 16-33B. From these data, draw the prophage map.

TABLE 16-33A

Parental phage	Percent wild-type recombinants
a^-b^-	12
a^-c^-	6
a^-d^-	1
b^-c^-	6
b^-d^-	11
c^-d^-	5

(c) What is the most likely physical structure of the DNA isolated from the phage (that is, linear, circular, supercoiled) and, if linear, what is the nature of the ends? Propose a physical experiment with isolated DNA to test your hypothesis.

TABLE 16-33B

Initial prophage genotype	Percent wild-type among lac^+ transductants
$a^-b^+c^+d^+$	20
$a^+b^-c^+d^+$	30
$a^+b^+c^-d^+$	40
$a^+b^+c^+d^-$	16

16-34   What are the differences between a phage that produces only specialized transducing particles, one that produces only generalized transducing particles, and one that can produce both?

16-35   Why are λ specialized-transducing particles generated only by induction rather than by lytic infection?

16-36   Some temperate phages that integrate their DNA do not seem to form specialized transducing particles. Give several possible reasons for this observation.

16-37   Could either RNA or single-stranded DNA phages carry out specialized transduction?

16-38   In 1965, H. Ikeda and J. Tomizawa carried out the following experiments. *E. coli* cells, which are genetically unable to synthesize thymine, were fed 5-bromouracil to make their DNA "heavy" and then were infected with the generalized transducing phage P1 and incubated in a medium containing (1) thymine, (2) 5-bromouracil, or (3) thymine plus radioactive phosphorus until lysis occurred. Analysis of phage progeny by density-gradient centrifugation showed that transducing particles from the first two media have similar densities and that those from the third are not radioactive. From these observations, what do you conclude about the phage genetic material in transducing particles and the origin of transduced segments?

16-39   The DNA of *Salmonella* phage P22 is terminally redundant and permuted. P22 is capable of both generalized and specialized transduction. If grown on *Salmonella* carrying an antibiotic resistance–transfer factor (RTF), P22 can transduce tetracycline resistance *(tet-r)* from the RTF to a sensitive *(tet-s)* *Salmonella* strain. Since P22 DNA is not large enough to carry the entire RTF DNA, the *tet-r* allele must be incorporated into the host DNA in order to be replicated. Consider a transduction experiment in which a lysate prepared on a *tet-r* strain is used to infect a *tet-s* cell which is lysogenic for P22. In general, the *tet-r* allele is located at or adjacent to the site of the original P22 prophage.

When transductants (*tet-r* P22 lysogens) are induced by ultraviolet light, the cells lyse and release a normal number of phage-like particles, which can be counted by electron microscopy. These lysates have several interesting properties: (1) Only one in 10^5 particles can form a plaque; (2) when these phage-like particles infect a *tet-s* *Salmonella*, they transduce the *tet-r* allele at a frequency of one transductant per ten particles; (3) when the lysate is used to infect *Salmonella* at various multiplicities of infection (moi), the number of particles produced (counted by electron microscopy) depends strongly on moi, that is, at moi = 0.03, 0.3, 3,; and 10, the numbers of phage produced per cell are 0.2, 20, 250 and 270, respectively. What is the nature of these phage-like particles? Hint: P22 packages DNA by the "headful" mechanism—that is, the length of DNA required to fill a phage head is cut from a concatemeric phage DNA molecule.

16-40   Following are some facts about specialized transduction by phage P22 in *Salmonella typhimurium*: (1) Specialized transducing particles of P22 carrying the *proA* and *proB* genes can be generated. These genes are immediately adjacent to the prophage attachment site on the bacterial chromosome. The *proA*⁻ and *proB*⁻ specialized transducing particles have the behavior described below. (2) Lysates of the specialized transducing particles can go through the lytic cyle, but only in mixed infections. (3) In single infections the particles transduce by substitution; however, in mixed infections, the particles transduce by lysogenization. What is the molecular basis of each of the foregoing observations?

17

Plasmids

●17-1 What is meant by a plasmid?

17-2 Could you have a plasmid with no genes whatsoever?

17-3 How do the sizes of plasmids compare with the size of phage DNA molecules?

17-4 (a) What is a cleared lysate?
(b) Can any plasmid be found in a cleared lysate?
(c) How are plasmids purified from a cleared lysate?

17-5 How is the migration rate of plasmid DNA in gel electrophoresis related to molecular weight and shape (supercoiled versus nicked circle)?

17-6 What is a relaxation complex?

17-7 Lac$^+$ and Lac$^-$ cells form purple and white colonies respectively on EMB lactose agar. A culture consisting of cells whose genotype is $F'lac^+/lac^-$ is mutagenized and plated on EMB-lactose agar at 30°C. Several purple colonies whose color is a little less intense than the others are studied further. These give white colonies at 42°C and light purple colonies at 30°C. Bacteria obtained from white (42°C) colonies remain white when replated at 30°C. What type of mutation has been acquired?

●17-8   (a) Define the following terms: conjugative, mobilizable, self-transmissible.
(b) Which of these terms describes F?
(c) Which of these terms describes ColE1?

17-9   (a) An *E. coli* cell containing F can transfer the chromosome at a frequency of about one transfer event per cell. Is this donation or conduction, and does it occur in a recombination-deficient cell?
(b) Can ColE1 cause chromosome transfer?

17-10   (a) What mode of DNA replication is used in transfer?
(b) What early enzymatic step is needed in transfer replication, but not in normal replication?
(c) If, in a particular cell type, rifampicin were to inhibit DNA transfer, what would you conclude about in the transfer mechanism?

17-11   What are the roles of DNA synthesis in the donor and the recipient?

17-12   What procedure can be used to transfer a nontransmissible plasmid from one strain to another?

17-13   Replication of unintegrated F, but not integrated F, is inhibited by exposing *E. coli* to acridine orange. Make use of this finding
(a) To obtain F^- from F^+ cells.
(b) To identify colonies as F^+, Hfr, or F^-.

17-14   Although the concentration of enzymes made is tightly regulated, it is often found that more enzyme is made per cell if an operon is located on an F' sex plasmid than if it is located on a bacterial chromosome. Explain.

17-15   What is the role of the sex pilus in conjugation?

17-16   Would you expect an Hfr strain of *E. coli* to have sex pili?

17-17   Pair formation occurs between an F^+ and an F^- cell yet does not occur between two F^+ cells. Why not?

17-18   In general an Hfr cell and an F^+ cell cannot mate because pair formation does not occur. Growth of a male cell (Hfr or F^+) in anaerobic conditions for a period of time converts it to a "phenotypic female"—that is, it lacks pili and can pair with a normally grown male. It is possible by the detection of genetic recombinants to observe the transfer of DNA from an Hfr to an F^+ phenotypic female. However, no genetic test has demonstrated transfer of the F plasmid to an Hfr phenotypic female. What is the cause of this asymmetry?

17-19   Integration of F to form an Hfr can often occur in two orientations at a single site. Since integration occurs by genetic recombination between homologous or nearly homologous sequences, what features must these sites have?

17-20   A Lac$^-$ bacterial strain has a $dnaA$(Ts) mutation so that it is unable to form a colony at 42°C. An $F'lac^+$ plasmid is introduced into the strain by conjugation. The culture is grown for many generations at 30°C, and then 10^6 cells are plated at 42°C. A few colonies arise and these are capable of growth at both 30°C and 42°C.
(a) Are these colonies Lac$^+$ or Lac$^-$?
(b) Do the cells still carry the $dnaA$(Ts) mutation?
(c) What feature of the cell has changed that enables them to grow at 42°C?
(d) Can the cells grow at 42°C in the presence of acridine orange?

17-21   An F'(Ts)lac^+ plasmid has a temperature-sensitive mutation in its replication system.
(a) What is the phenotype of an F'(Ts)lac^+/lac^- cell at 42°C?
(b) An F'(Ts)lac^+/lac^-gal^+ strain is grown for many generations and then plated at 42°C. Some Lac$^+$ colonies form at 42°C. How have these formed?
(c) Some of the Lac$^+$ colonies in (b) are Gal$^-$. How have these formed?

17-22   A strain carrying $F'gal^+$, which forms red colonies in MacConkey-galactose agar (Gal$^-$ colonies are white), is mutagenized and plated. A few colonies are found that are slightly smaller and more intensely red. Further study shows that they have ten copies of $F'gal^+$ per cell rather than the usual number. What types of mutations have occurred?

17-23 Plasmids contained in Rec^+ cells frequently dimerize. A dimer can then be isolated and transferred by the $CaCl_2$ technique to a Rec^- cell in which it will be stable and not revert to the monomer. If a monomer plasmid has a copy number of ten, what will the copy number of the dimer be?

17-24 On EMB-lactose agar Lac^+ and Lac^- colonies are purple and white respectively. If several thousand $F'lac^+/lac^-$ cells are plated, a few sectored colonies appear. This may have a purple half and a white half or a wedge of white in a predominately purple colony. Sectored colonies that are predominately white are not found. Explain the cause of sectoring.

17-25 Complete transfer of the chromosome in Hfr transfer requires about 100 minutes.
(a) At what time(s) is (are) F DNA transferred?

(b) A Str-s Hfr that transfers the genes in alphabetical order, a b ... y z, is mixed with a F^-Z^-Str-r cell. The mixture is agitated violently 15 minutes after mixing to break apart conjugating cells and then plated on agar lacking Z and containing streptomycin. About one cell in 10^7 yields a Z^+Str-r colony. What is the genotype of such a colony?

17-26 With certain combinations of donor and recipient cells plasmid transfer is very inefficient. For example, if two *E. coli* strain K cells (one $F'lac^+$ and one F^-lac^-) are mated for 15 minutes, the number of F^-lac^- cells converted to Lac^+ is roughly equal to the number of $F'lac^+$ cells. However, if the F' is contained in strain K and the F^- cell is strain B, the number is about 10^5-fold smaller. These strain B $F'lac^+$ cells can then be used in subsequent matings. When this is done, it is found that in a B x B mating, the transferred plasmid is detected in both F^- cells, whereas in a B x K mating, the frequency is again quite low. Explain this phenomenon.

18

Homologous Recombination

•18-1 In its most general meaning what is meant by genetic recombination?

•18-2 Distinguish homologous recombination, transposition, and illegitimate recombination with respect to a requirement for homology and for RecA function.

•18-3 (a) What is a heterozygote, in genetic terms?
(b) Give two molecular arrangements that could give rise to genetic heterozygosity in a haploid organism.

18-4 Which of the following are examples of reciprocal exchanges in genetic recombination? Prophage integration; prophage excision; formation of recombinants in an Hfr x F^- bacterial cross; generalized transduction; formation of an F' factor from an Hfr bacterium; formation of an Hfr bacterium from an F^+ cell; bacterial transformation.

18-5 Suppose that phage mutants are isolated which show decreased ability to recombine. All the mutants fall into two complementation groups, recX and recY. The genes are in the following map order: *d e f g h j k l m n.* Each of the ten genes is one recombination unit apart (that is, $d \times e = 1$ percent, $j \times k = 1$ percent, and so on). Crosses are now performed with Rec⁻ mutants. The data in Table 18-5 are obtained.

(a) What properties of the *recX* and *recY* systems are indicated by these results?

(b) What is the probable relation between the physical spacing of adjacent genes?

TABLE 18-5

		Recombination frequency, percent	
Genotypes of parental phages in crosses	Genotype of recombinant	recX-phages	recY-phages
d^-, e^-	$d^+ e^+$	0.0001	1
f^-, j^-	$f^+ j^+$	0.0001	4
j^-, m^-	$j^+ m^+$	1	2
j^-, n^-	$j^+ n^+$	1	3
k^-, m^-	$k^+ m^+$	1	1
k^-, l^-	$k^+ l^+$	1	0.01
j^-, k^-	$j^+ k^+$	0.0001	1
l^-, m^-	$l^+ m^+$	0.0001	1

18-6 *E. coli* Q is *recA⁻* (that is, it is deficient in bacterial recombination) and contains a portion of a λ prophage—only the *J* gene and the right prophage-attachment site (see the λ map, Appendix C). The *J* gene is mutant and codes for the *h* character. Wild-type λ cannot adsorb to the λ-resistant strain K/λ, but a λ mutant carrying the *h* marker can. If a *cI⁻* mutant of λ is plated on a bacterial lawn consisting of a 1:3 mixture of Q and K/λ, plaques result which are easily seen, although they are somewhat turbid. Many phage recovered from the plaques can form plaques on K/λ. When a mutagenized stock of λ*cI⁻* is plated on this mixture, some plaques are found which are so turbid that they are almost not visible. These plaques do not contain phage that can form plaques on K/λ. What is the probable genotype of the phage producing these very turbid plaques?

18-7 Two homologous DNA structures are shown diagrammatially. In each molecule a + strand is complementary to its own - strand and to the - strand of the other molecule. The molecules are oriented so that the b ends are homologous in base sequence. The upper molecule is shorter than the lower one and lacks the a end. (Figure 18-7). The molecules are allowed to hybridize and to branch-migrate. Which of configurations—(1) to (6)—could result from the operation of these two processes?

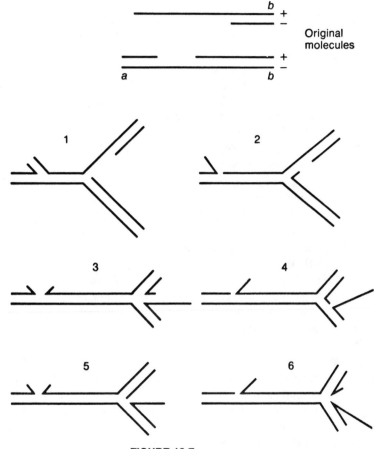

FIGURE 18-7

18-8   Radding did an experiment in which both superhelical and nicked circles of the replicative form of the phage øX174 were separately mixed with single-stranded fragments obtained from the DNA of the phage. No RecA protein was added. At a temperature somewhat above room temperature strand invasion occurred in one of these molecules, and it was not possible to find a temperature at which it would occur in both molecules. Explain these results.

18-9   List the two biochemical activities of the RecA protein.

18-10   Ultraviolet irradiation of phage stimulates genetic recombination
        between phage DNA molecules in an infected cell. This stimulation
        rarely occurs unless the host possesses an active Uvr repair
        system. Suggest a possible basis for this stimulation.

18-11   The recombination frequency for phage Q is generally very low.
        Using genetic markers that allow identification of a heterozygous
        particle, a single burst experiment is done.
        (a) 90 percent of the bursts contain no recombinants or het-
        erozygotes, 5 percent have a single heterozygote and no re-
        combinants, and 5 percent contain one or two recombinants but no
        heterozygotes. At what stage in the life cycle of the phage do
        you think recombination occurs?
        (b) In what fraction of the infected cells is there re-
        combination? Explain why your estimate is a minimum number.
        (c) Another phage W yields different results, namely, 99 percent
        of the bursts contain neither recombinants nor heterozygotes, and
        1 percent of the bursts contain phage nearly half of which are
        recombinant. No heterozygotes are seen. What can you say about
        the timing and frequency of recombination?
        (d) With another phage Y, 95 percent of the bursts contain
        neither recombinants nor heterozygotes and 5 percent of the
        bursts contain three or four heterogygotes and many recombinants.
        What can you say about the timing and frequency of recombination
        with this phage?

18-12   In the simplest break-rejoin models of recombination the initial
        exchange must occur between the genetic markers that have been
        recombined. In the strand transfer model this is not necessary.
        What feature(s) of the model eliminate(s) this necessity?

18-13   What physical features define a Holliday intermediate?

18-14   In studying figure-8 molecules of phage ϕX174 by electron
        microscopy it is important to be sure that an observed figure-8
        did not simply arise by two circular molecules overlapping one
        another.
        (a) What criteria must be applied to an observed structure before
        even considering that it might be a figure-8?
        (b) What could you do when preparing the DNA sample for electron
        microscopy to avoid the overlap problem?

18-15 Figure-8 molecules have been detected by electron microscopy by converting them to X molecules with a restriction enzyme that makes a single cut in each circle.

(a) Is the observed junction probably at the same position as it was in the figure-8 molecule at the time of isolation from infected bacteria?

(b) Figure-8 molecules are quite stable but if X molecules are stored for several hours before preparing them for electron microscopy, no X molecules are found—only linear molecules are seen. Explain this phenomenon.

18-16 The following experiment was done in the laboratory of F. W. Stahl. A λ phage with genotype $red^-gam^-A^-$ was density labeled with ^{13}C and ^{15}N and was mixed with $\lambda red^-gam^-R^-$ containing ^{12}C and ^{14}N. The mixture was used to infect E. coli using conditions in which DNA replication was inhibited and in which each bacterium was infected by ten phage particles. The phage that were produced were centrifuged to equilibrium in Cs formate and the distribution of genotypes throughout the density gradient was determined. The distribution of total phage and A^+R^+ recombinants shown in Figure 18-16 was observed. In a second experiment both

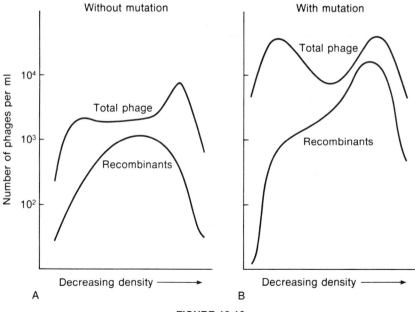

FIGURE 18-16

phage contained a mutation known to map approximately 93 percent from the left end of the phage and the distribution shown in Figure 18-16 was observed. Explain both results. The genetic map of phage λ is in Appendix C.

18-17   A phage has a gene a and a gene x at opposite ends of its linear DNA molecule. A bacterial culture is infected with equal numbers of a^- and x^- phage particles using conditions in which DNA replication cannot occur. The average burst size of the phage is one and 25 percent of the progeny have the genotype a^+x^+. What fraction of the input DNA molecules have engaged in recombination?

19

Transposons

●19-1 In homologous recombination two DNA molecules break and rejoin. If the event is physically reciprocal, the amount of DNA present is the same as that present before recombination; if the event is nonreciprocal, the total amount of DNA present decreases. What can be said about the relative amounts of DNA before and after transposition?

●19-2 When transposition occurs, two base sequences are duplicated. What are they?

●19-3 What is meant by the terms "direct repeat" and "inverted repeat"? Use the sequence ABCD as an example.

19-4 What is usually meant by an insertion or IS element?

●19-5 Since there is at present (1982) no detailed understanding of the mechanism of transposition, what is the evidence that compels one to conclude that DNA replication is an essential step?

19-6 It is easy to understand how insertion of a transposon can cause a mutation, but what feature of some transposons is responsible for genetic polarity?

19-7   You have isolated a segment of DNA having about 500 base pairs. Several genetic experiments have suggested but have not proved that the segment contains a transposon. What physical or chemical experiment would you do to test the hypothesis?

•19-8 Are all drug-resistance genes located in transposons?

19-9   An Amp-r plasmid whose replication is temperature-sensitive is introduced into an Amp-s cell by the $CaCl_2$-transformation method. After grown for several generations 10^7 cells are plated on agar containing ampicillin at 42°C. Fifty colonies form.
       (a) Give three mechanisms which could explain the presence of these colonies.
       (b) Which mechanisms would not occur in a RecA$^-$ cell?

19-10 You have isolated two different insertion mutants in phage λ. How would you determine whether the two insertions are the same transposon?

19-11 A particular IS element has inserted into different positions in two λ phages. In one case the insertion is at position 55; in the other, at position 75 (both measured on a scale of 100 from the left end of the phage. The IS element has a length equal to 1 percent of the size of λ DNA. One of these phage variants shows polar effects on downstream genes; the other does not.
       (a) Explain the difference in the polar effects.
       (b) How could your explanation best be tested by electron microscopy?

19-12 Design a simple procedure for isolating the DNA of a transposon.

19-13 Sometimes in a bacterial strain carrying two different plasmids, one of which contains a transposon, a single plasmid arises that has the genetic information of both plasmids. If the cell is RecA$^-$, this fusion can occur in two possible ways—(1) illegitimate recombination and transposon-mediated formation of a cointegrate or (2) some other form of nonhomologous recombination.
       (a) What information would be necessary to distinguish these mechanisms unambiguously?
       (b) What physical experiment could you do to obtain this information?
       (c) What biological experiment might you do to obtain this information?

19-14 Two bacterial genes (*chk* and *lat*) are very near one another. When *chk⁻* mutants are isolated (they arise spontaneously at a frequency of about 10^{-6} mutants per cell per generation), about 1 percent of these mutants are also *lat⁻*. This is a surprisingly high frequency for spontaneous formation of double mutants.

(a) Could you guess what might be the cause of this phenomenon?

(b) Would a measurement of this frequency in a RecA⁻ cell be informative? Explain.

(c) Would studying the effect of chemical mutagens be informative? Explain.

19-15 The frequency of reversion of transposon-induced polar mutations is very low. The frequency is higher in a RecA⁺ cell than a RecA⁻ cell, but the reason for this has not been unambiguously determined. Can you suggest a reason?

19-16 It is sometimes desirable to remove a transposon from a cell. This is possible because excision of transposons occurs spontaneously, though at a very low frequency and by an unknown mechanism.) If a transposon were inserted in the *lac* gene, one would only have to obtain a Lac⁺ revertant colony to obtain cells that had lost the transposon. In some genes, revertants are not detected in such a straightforward way; however, the following technique is successful for transposons carrying some antibiotic-resistance markers. Consider a culture of cells in which a transposon carrying the *tet⁺* gene is present in gene *x* rendering the cells X⁻. Tetracycline is added to a growing culture and a few minutes later penicillin is added. After many hours of growth the cells are plated on agar lacking tetracycline. At a low but easily detected frequency colonies form whose cells lack the transposon.

(a) What principle underlies this technique?

(b) Describe three mechanisms for transposon loss in these cells.

19-17 Since transposons are duplicated at a reasonably constant rate, one might expect that in the course of the millions of generations experienced by each bacterial species, a particular bacterial cell would contain thousands of copies of any transposon that it may have acquired in the distant past. However, one or two copies is the norm.

(a) Explain this low number.

(b) Suppose in nature a mutation had occurred which increased the transposition frequency 1000-fold compared to the frequencies usually observed. Would you expect such a transposon to exist today?

(c) If the mutation had occurred in a laboratory strain, would you expect the situation to be different from that in part (b)?

20

Recombinant DNA and Genetic Engineering

●20-1 Define cloning, vector, and vehicle as they are used in the recombinant DNA technology.

●20-2 These questions concern restriction endonucleases.
(a) What is meant by a restriction enzyme?
(b) What do you think is the biological role of bacterial restriction enzymes?
(c) Why do bacteria not destroy their own DNA by their restriction enzymes?
(d) What is a type I and a type II enzyme?
(e) What common feature is present in every base sequence recognized by a restriction enzyme?

● 20-3 Two types of cuts are made by restriction enzymes. What are they and what terms are used to describe the termini formed?

20-4 What property must two different restriction enzymes possess if they yield identical patterns of breakage?

20-5 Restriction enzymes A and B are used to cut the DNA of phages T5 and T7, respectively. A particular T5 fragment is mixed with a particular T7 fragment and a circle forms that can be sealed by DNA ligase. What can you conclude about these enzymes?

20-6   T7 DNA is digested with the EcoRI enzyme. Can all fragments circularize?

20-7   Is it possible for two restriction enzymes (acting separately on two DNA samples) to yield sets of fragments that are different but have one fragment whose size (number of nucleotides) in one set is identical to the size of a fragment in the second set?

●20-8   What properties are required to make a plasmid suitable as a cloning vehicle?

●20-9   How is recombinant plasmid DNA introduced into bacterial cells in which it can replicate?

20-10   Plasmid pBR607 DNA is circular and double-stranded and has a molecular weight of 2.6×10^6. This plasmid carries two genes whose protein products confer resistance to tetracycline (Tet-r) and ampicillin (Amp-r) in host bacteria. The DNA has a single site for each of the following restriction enzymes: EcoRI, BamHI, HindIII, PstI, and SalI. Cloning DNA into the EcoRI site does not affect resistance to either drug. Cloning DNA into the BamHI, HindIII and SalI sites abolishes tetracycline-resistance. Cloning into the PstI site abolishes ampicillin-resistance. Digestion with the following mixtures of restriction enzymes yields fragments with the sizes listed in Table 20-10. Position the PstI, BamHI. HindIII, and SalI cleavage sites on a restriction map, relative to the EcoRI cleavage site.

TABLE 20-10

Enzymes in mixture	Molecular weights of fragments (millions)
EcoRI, PstI	0.46, 2.14
EcoRI, BamHI	0.2, 2.4
EcoRI, HindIII	0.05, 2.55
EcoRI, SalI	0.55, 2.05
EcoRI, BamHI, PstI	0.2, 0.46, 1.94

20-11 The DNA of the plasmid pHUB1 is circular, double-stranded, and contains 5.7 kilobase pairs (5.7 kb). This plasmid carries a gene whose protein product confers resistance to tetracycline (Tet-r) on the host bacterium. The DNA has one site each for the following restriction enzymes: EcoRI, HpaI, BamHI, PstI, SalI, and BglII. Cloning into the BamHI and SalI sites abolishes tetracycline-resistance; cloning into the other sites does not. Digestion with various restriction enzymes and combinations of enzymes yields DNA fragments whose sizes in kilobase pairs are shown in Table 20-11. Position the cleavage sites for all these enzymes on a map of the pHUB1 DNA. Draw the map as a circle with 5.7 kb marked off on the circumference.

TABLE 20-11

Enzymes in mixture	Fragment size (kilobases)
EcoRI	5.7
EcoRI, BamHI	0.4, 5.3
EcoRI, HpaI	0.5, 5.2
EcoRI, SalI	0.7, 5.0
EcoRI, BglII	1.1, 4.6
EcoRI, PstI	2.4, 3.3
PstI, BglII	1.3, 4.4
BglII, HpaI	0.6, 5.1

20-12 The enzyme terminal transferase is a DNA polymerase in the sense that it adds a nucleotide 5'-P to a free 3'-OH group. What property does it have that the usual polymerases do not have? Explain how terminal transferase can be used to join together two DNA molecules.

20-13 What two methods can be used to join blunt-ended fragments?

20-14 In analyzing DNA by treatment with a restriction endonuclease followed by gel electrophoresis, in a particular experiment it is observed that in addition to the expected bands, DNA is present in all regions of the gel starting from the position of the most slowly-moving fragment and past that of the most rapidly-moving fragment. If the DNA is analyzed before treatment with the restriction enzyme, no such material is seen. Suggest an explanation.

20-15   When DNA is analyzed by treatment with a restriction endonuclease followed by gel electrophoresis, it is almost always necessary to heat the mixture to 65°C before layering it on the gel. If this is not done, extra bands are seen. Why is this the case?

20-16   A large circular DNA plasmid is digested with a restriction endonuclease. The fragments are allowed to reassemble at random (that is, the fragments are no longer in the original order) and circles having the same molecular weight as the original plasmid are selected. Host cells lacking the plasmid are then infected with these fragments.
        (a) Assuming that ability to replicate is the only requirement for successful infection, what alterations in fragment order might prevent successful infection?
        (b) Consider a larger circle in which the order of the fragments is the same, but in which some circles contain an additional fragment adjacent to an identical fragment. Will this circle be capable of successful infection?
        (c) Is it possible to delete a fragment, keeping the order otherwise the same, and still infect cells successfully? Explain.

20-17   A restriction digest is annealed at 45°C, cooled, treated with ligase, and electrophoresed. The number of bands observed increases with the total DNA concentration and the time of annealing. Explain this phenomenon.

20-18   A DNA molecule has the restriction map shown in Figure 20-18 when digested by the EcoRI restriction endonuclease (numbers refer to relative distance of each cut from the left end of the molecule). In a particular experiment (after heating the digested sample to 65°C), gel electrophoresis shows that instead of the expected five bands, the second pattern shown is observed. Explain.

20-19   A phage DNA molecule has short, complementary, single-stranded ends and circularizes when the phage infects a bacterium. The restriction map and the gel band pattern obtained after treatment of a free DNA molecule with the Bam restriction endonuclease are shown at A and B in Figure 20-19. You are studying the properties of DNA isolated from an infected bacterium. Under certain conditions the pattern of fragments obtained after enzymatic

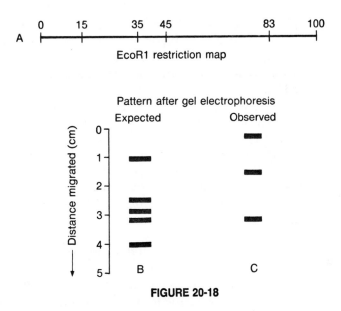

FIGURE 20-18

digestion and gel electrophoresis is that shown in panel C. What is the structure of the DNA under these conditions?

20-20 The DNA of temperate phage P4 is linear, double-stranded, 11.5-kb long, and has cohesive ends. Digestion with the Bam restriction

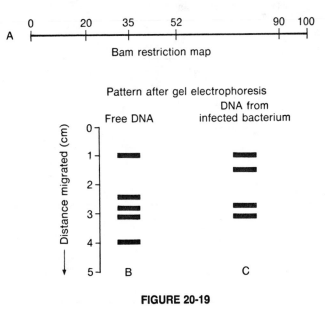

FIGURE 20-19

endonuclease yields fragments 6.4, 4.1 and 1.0 kb in length. Partial digestion with the Bam enzyme yields fragments 10.5, 7.4, 6.4, 4.1, and 1.0 kb in length. Circular P4 DNA made with DNA ligase can be digested with the Bam enzyme to yield fragments 6.4 and 5.1 kb in length.

(a) What is the order of the fragments in the DNA?

(b) DNA is extracted from cells lysogenic for P4 and digested completely with the Bam enzyme. The fragments are separated by gel electrophoresis, transferred to nitrocellulose paper, and hybridized with radioactive P4 phage DNA. Fragments whose lengths are 15.0, 12.5, and 6.4 kb are labeled with radioactivity. What conclusions can you draw from this experiment about prophage structure?

20-21   Wild-type phage X4 DNA is treated with a restriction enzyme and five fragments—I, II, III, IV, and V—are separated and purified by gel electrophoresis. It has been decided to map these fragments by comparison to the genetic map of X4. This is done by performing twenty-five transfection experiments using the five mutations shown in Figure 20-21. The experiment consists of transfection of a Rec+ cell with intact DNA from a single mutant and a purified fragment. The data are shown in Table 20-21. (The values indicate the number of plaques per microgram of intact DNA molecules.) What is the order of the fragments starting at the left end of the map?

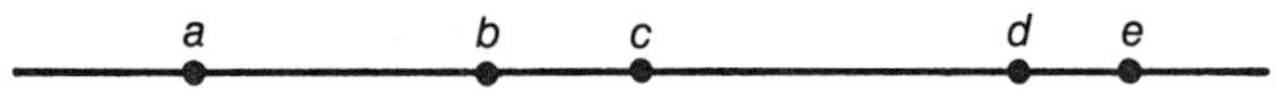

FIGURE 20-21

TABLE 20-21

	Fragments				
Mutant	I	II	III	IV	V
a^-	21	14	21,842	3	8
b^-	6	0	4	4,856	2
c^-	32,681	16	8	11	13
d^-	1	3	0	1	1,825
e^-	22	17,813	27	18	30

20-22 A Tet-r plasmid having a single EcoRI site is cleaved by the EcoRI enzyme. The sample is annealed, treated with DNA ligase and then used in a CaCl² transformation experiment. Tet-r colonies are selected. Gel electrophoresis of the plasmid DNA from these colonies shows that there are two types of colonies—one type yielding a single band at the expected position for the original plasmid and the other type yielding a band moving much more slowly. What might the slow band represent?

20-23 A Kan-r Amp-r plasmid is treated with the BglI enzyme which cleaves the *amp* gene. The DNA is annealed with a BglI digest of *Drosophila* DNA and then used to transform *E. coli*.
(a) What antibiotic would you put in the agar to insure that a colony has the plasmid?
(b) What antibiotic-resistance phenotypes will be found on the plate?
(c) Which phenotype will have the *Drosophila* DNA?

20-24 The 16S and 23S ribosomal RNA species originate from a single 30S precursor RNA. The DNA segment from which this 30S rRNA is transcribed is called rrn. It is known that rrn is redundant in *E. coli*; the number of rrn copies has been established on the basis of saturation hybridization as 5 to 10. Specialized transducing particles carrying rrn have been isolated carrying DNA segments from positions 71-min, 83-min, 85-min, 88-min, and 89-min on the map of the *E. coli* chromosome, suggesting the existence of at least five rrn "genes." The rrn genes are probably all identical, since rRNA from *E. coli* is homogenous, and since restriction analysis of the DNA within the rrn genes at all five of the above-mentioned loci gives the following result: There are no Bam sites in the rrn segment of these phages, but there are two Sal sites, one within the 16S coding portion and one within the 23S section, as shown below. In order to find out the number of rrn in *E. coli*, either *E. coli* DNA or the DNA isolated from λ transducing particles carrying rrn DNA was completely digested with Bam or Sal restriction nucleases, and fragments were separated by electrophoresis. The bands were transferred to nitrocellulose paper and either radioactive 16S or 23S rRNA was hybridized to the paper. The paper was washed and placed in contact with photographic film. The developed film, in which only radioactivity produces blackening, is shown in Figure 20-24B. Judging from these data, how many rrn genes must there be in *E. coli*? (Figures on page 170.)

FIGURE 20-24A

DNA	λrrn	E.coli	λrrn	E.coli	λrrn	E.coli	λrrn	E.coli
Restriction nuclease	Bam	Bam	Bam	Bam	Sal	Sal	Sal	Sal
^{32}P-RNA	16S	16S	23S	23S	16S	16S	23S	23S

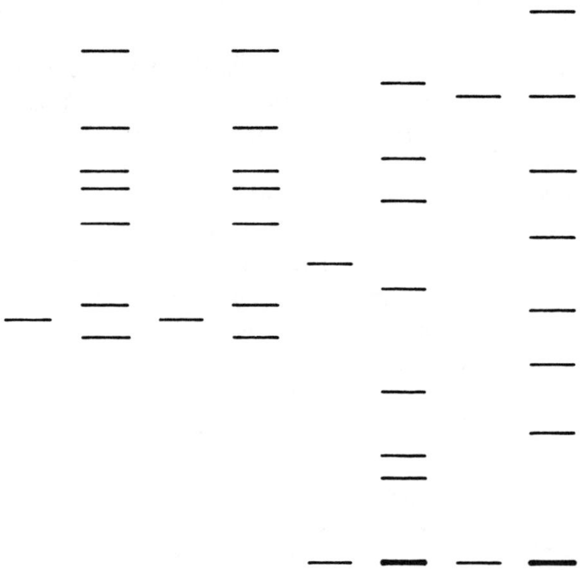

FIGURE 20-24B

20-25   The following elegant method has been used to obtain the
restriction maps of a DNA molecule for different enzymes. First,
the 5' termini of a phage are labeled with ^{32}P using polynucleotide kinase. The sample is divided in half and each part is
digested either with enzyme I or enzyme II. The two digests are
separately electrophoresed and the radioactive bands are noted.

This experiment identifies the terminal fragments, information that is used later. The major part of the technique uses both an unlabeled and a uniformly ^{32}P-labeled sample of the DNA. The unlabeled sample is digested with enzyme I and electrophoresed using an arrangement that gives broad bands. These bands are located by the standard technique of ethidium bromide fluorescence, their positions are recorded, the DNA is denatured by immersing the gel in alkali, and then the bands are transferred to a nitrocellulose sheet to yield the configuration shown in Figure 20-25. The ^{32}P-labeled sample is treated with enzyme II, electrophoresed, the positions of the bands are noted, the DNA is denatured wih alkali, and then transferred to the nitrocellulose paper containing the enzyme I fragments. The transfer is done so that the ^{32}P bands are at right angles to the unlabeled bands, as shown in the figure. The paper is then

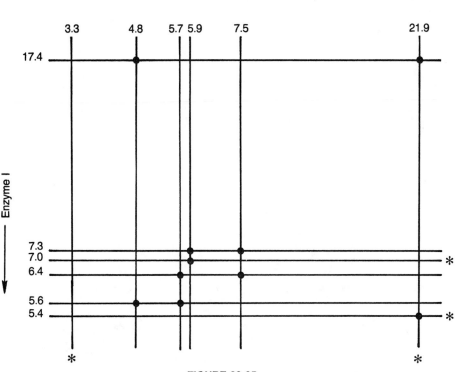

FIGURE 20-25

subjected to renaturing conditions, washed, and the radioactivity is located by autoradiography. Figure 20-25 shows the positions of the bands from each gel (thin lines) and the positions of the radioactivity (small circles). The bands marked with asterisks are the terminal fragments. What are the restriction maps for each enzyme? The numbers indicate the size of each fragment in kilobase pairs and the solid circles indicate radioactive regions.

20-26 How many clones are needed to establish a gene library for monkey DNA (molecular weight = 1.74×10^{12}) using fragments whose molecular weights average 10^7, if one wishes 99 percent of the monkey genes to be in the library?

21

Animal and Plant Viruses

● 21-1   What kinds of genetic material are present in different viruses?

● 21-2   What are (-)-strand RNA viruses?

   21-3   (a) What distinguishes naked from enveloped viruses?
           (b) What are the common basic symmetries of naked viruses?

●21-4   (a) What is meant by the terms "early" and "late" when applied to virus-coded functions or proteins?
           (b) What kinds of functions are frequently found to be early?
           (c) What kinds are late?

   21-5   What kinds of viruses must carry enzymes for mRNA synthesis as part of the virion and introduce them into the cell upon infection?

   21-6   What kinds of viruses must carry enzymes for DNA synthesis as part of the virion and introduce them into the cell upon infection?

●21-7   Define the terms capsid, protomer, capsomer, hexon, penton, and virion.

21-8   Some viruses contain substantial quantities of phospholipids, cholesterol, and carbohydrates. What is the source of these substances?

• 21-9   Define segmented genome, heterocapsidic virus, associated virus, and covirus.

21-10   (a) What features do (+)-strand viruses share with eukaryotic mRNA?
(b) With any single-stranded RNA virus, which strand has a ribosomal binding site?

21-11   A virus sample is suspended in 3 M potassium tartrate and then centrifuged. After 6 hours of centrifugation it is observed that there is no pellet at the bottom of the centrifuge tube but, instead, there is a thin translucent film floating on the surface of the liquid. In another experiment the virus sample is exposed to ether and then centrifuged in the same way; after centrifugation a pellet is found and no material is on the surface of the liquid. Explain what is happening and what might be the composition of the virus.

21-12   (a) What are the three major mechanisms for release of progeny virions by an infected cell?
(b) Which ones are used by naked and by enveloped viruses?

21-13   Efficiency of plating (EOP) is defined as the fraction of particles that can form a visible plaque. With phages, the EOP is generally nearly one, whereas with viruses, values of from 10^{-6} to 10^{-1} are more common. What factors contribute to the very low values of EOP of viruses?

• 21-14 RNA viruses have four distinct mechanisms for producing mRNA. What are these mechanisms?

• 21-15 Viruses must synthesize several proteins and must deal with the problem that in eukaryotes only a single protein can be translated from an mRNA molecule. What mechanisms are used to circumvent their problem?

• 21-16 The total molecular weight of several animal viruses exceeds the coding capacity of the virus, if a single reading frame is used. Three major mechanisms contribute to this phenomenon. What are they?

21-17 A pulse-chase experiment in virus-infected cells is performed with ^{3}H-leucine. At various times the molecular weights of the labeled proteins are determined using SDS-PAGE electrophoresis, which employs conditions that denature the proteins and dissociate subunits. What can you conclude about particular viral systems in which the following observations are made?

(a) If the labeling is done between two and three hours after infection, label is present at four hours but not at twelve hours?

(b) One labeled band is found three hours later and six bands are found eight hours later. The molecular weights of the six proteins are smaller than those of the proteins at three hours postinfection.

(c) One labeled band (molecular weight = 6×10^5) is seen at early times, three (molecular weights = 5.6×10^5, 25×10^3, and 15×10^3) at a later time, and seven (molecular weights = 11×10^4, 8×10^4, 7×10^4, 6×10^4, 5×10^6, 4×10^6, and 3×10^6) at very late times.

21-18 What mechanism ensures that each reovirus particle gets one copy of each RNA molecule of its segmented genome?

21-19 (a) How many components are there in Rous sarcoma virus RNA and what are they?

(b) What is the function of each?

21-20 What is the difference between the way that integration fits into the life cycles of retroviruses and the life cycles of the DNA tumor viruses?

21-21 Certain features of the life cycles of viruses are analogous to those in phage-infected bacteria—for example, establishment of new replication and transcription systems, synthesis of huge amounts of structural proteins that self-assemble, and except for the simplest viruses, distinct periods of early and late transcription. However, certain features of viral systems are rarely found, if ever, in phage systems. Name some of these features.

●21-22 What forms of nucleic acid are isolated from polyoma virus and from SV40 virus?

●21-23 What is T antigen?

21-24   (a) What is the Hirt procedure?
        (b) What criterion might you use to determine whether all of
        the replicating DNA is recovered by this procedure?

21-25   What mechanism does SV40 utilize that allows the total molecular
        weight of its protein to exceed the coding capacity of its
        DNA?

21-26   Neither SV40 nor polyoma virus encodes a DNA polymerase. What
        does this fact tell you about the cellular location for
        replication?

21-27   Adenovirus DNA molecules are able to circularize yet they do not
        have complementary single-stranded termini. How do they
        circularize?

21-28   Priming of DNA synthesis of adenovirus DNA does not require the
        activity of any known RNA polymerase. How does priming
        occur?

21-29   (a) What is the meaning of the terms "permissive" and
        "nonpermissive" in viral infections?
        (b) What is the distinction between productive infection and
        transformation by papovaviruses?

21-30   Explain the term contact inhibition or, as it is now more
        frequently called, density-dependent growth.

21-31   What is a transformed cell? Distinguish a transformed cell from a
        normal cell with respect to density-dependent growth, anchorage
        dependence, serum requirements, and lectin agglutination.

21-32   (a) How is cell fusion carried out?
        (b) If a cell with 22 chromosomes is fused to another species of
        cells with 30 chromosomes, what is the maximum and minimum number
        of chromosomes found in cells which are hybrid for some
        particular gene?

21-33   How can one prove that a given papovavirus-transformed cell line
        contains the entire viral genetic material?

21-34   Some years ago a considerable amount of thought was given to
        seeking parallels between lysogenic phages and tumor viruses and

phages such as λ were considered to be model systems for understanding virus-induced cancer. In recent years, the differences have become quite apparent and lysogeny is now thought to have little relation to tumorigenicity. List some of these differences.

21-35 What kind of viral mutant of a DNA virus might be able to transform a cell that is permissive for the wild-type virus?

22

Regulation of Eukaryotic Systems

●22-1 What is meant by a repetitive sequence, a unique sequence, and satellite DNA?

22-2 A Cot curve is shown in Figure 22-2 (page 179).
(a) What fraction of the DNA is contained in components that are unique? What fraction is redundant? What fraction is satellite?
(b) If the unique sequences are represented once per genome, how many copies per genome are there of each of the other two classes?

●22-3 (a) Why are eukaryotic genes not organized in operons?
(b) How does a gene family differ from an operon?

22-4 In discussing gene families one frequently mentions embryonic, fetal, and adult forms of a protein. To what phenomenon do these terms refer? What term is used to describe a gene family of this sort?

●22-5 (a) Are all members of a gene family transcribed in the same direction?

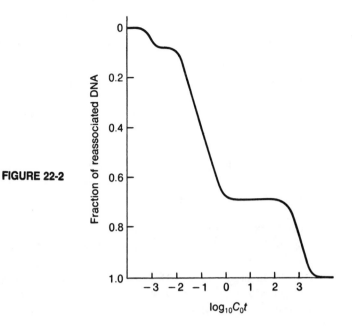

FIGURE 22-2

(b) Are all members of a gene family actively transcribed in all tissues of a single organism?

(c) Are all members of a gene family on the same chromosome?

(d) What is a fairly common location for repetitive sequences in clustered gene families?

22-6 (a) What is meant by totipotency?

(b) Differentiated cells rarely return to the undifferentiated state. Suggest several mechanisms for maintaining a differentiated state.

22-7 (a) What role does genetic recombination play in rDNA amplification in *Xenopus* oocytes?

(b) Several types of nucleases must be required for rDNA amplification. What are these types?

(c) The excess rDNA in *Xenopus* oocytes is not present in cells of an embryo. Explain what has happened to this DNA.

(d) The signals that cause the phenomenon in (c) are not known. Suggest a few possibilities.

22-8 Differentiated cells are often called upon to produce an enormous amount of a particular protein. What methods are used when only a limited time is available for this synthesis and when a long time is available?

22-9   Suggest a mechanism by which a cell might be signaled to synthesize a particular protein but no other.

• 22-10   Describe in outline the mechanism for generating antibody diversity.

22-11   What, in outline, is the mechanism that enables an organism to make a large amount of antibody to a particular antigen and to retain this programming indefinitely?

22-12   (a) If you wanted to redesign the mouse genome to increase antibody diversity, but by adding the least amount of DNA, which component would you increase?
(b) If an organism has 150 different *V* genes, 12 *J* genes, and 3 possible *V-J* joints for L chains, and can make 5000 different H chains, how many different antibodies can be made?

22-13   A particular hormone typically acts on only one or a small number of cell types. What is the usual property of a cell for determining whether a cell does or does not respond to a hormone?

• 22-14   Distinguish gene amplification from translational amplification.

22-15   A very abundant mRNA is isolated from a eukaryotic cell. You are interested in knowing how many copies of the corresponding gene there are per cell. You do this by denaturing and renaturing the total cellular DNA, obtaining a Cot curve. At various times of renaturation DNA samples are taken and you determine by DNA-RNA hybridization whether the gene has renatured yet. In this way, the position of the gene on the Cot curve is determined. From the abundance of the mRNA can you predict where on the Cot curve this DNA will be? Explain.

22-16   (a) Under what circumstances might the concentration of a particular protein be regulated by turning total protein synthesis on and off, as in the case of globin synthesis.
(b) Why can globin synthesis not be transcriptionally regulated?

22-17　A single amylase gene gives rise to two distinct mRNA molecules differing in their 5' untranslated regions but encoding the same protein. One form of mRNA predominates in salivary glands, where amylase mRNA comprises 2 percent of the total mRNA; the other predominates in liver and comprises only 0.2 percent of the mRNA. The relative concentration of amylase activity in the salivary gland and the liver is 4000:1. From what you have studied about translational control, suggest a way in which this difference in 5' ends of the mRNA molecules could exert control over levels of amylase protein synthesis.

22-18　When chromatin from somatic cells of X. laevis is isolated and transcribed in an *in vitro* extract, only somatic genes are transcribed. This is true whether the *in vitro* transcription extracts are derived from oocytes or from somatic cells. When the chromatin is washed with 0.6 M NaCl before transcription, this specificity is lost and oocyte genes are also transcribed. How do you interpret these results? What is the locus of specificity?

22-19　A particular cell makes substance X in response to a hormone H. Prior to exposure of a culture of the cells to H, a large RNA molecule can be found in the nucleus. It is capped and has a poly(A) tail. After addition of H, this RNA is not found but a second RNA, which in an *in vitro* protein-synthesizing system makes X, is found in cytoplasmic polysomes. Both RNA molecules hybridize to the same cloned DNA molecule. What simple mechanism might be used for hormonal control of synthesis of X?

22-20　A substance Q is made in response to an external effector E. In the absence of E, a nuclear RNA consisting of about 800 nucleotides can be found that hybridizes with the cloned gene encoding Q. The RNA is capped, has a poly(A) tail and in an *in vitro* protein-synthesizing system directs synthesis of a tri-peptide. If E is added, very little of this nuclear RNA is present and a mRNA molecule containing 770 nucleotides is found in the cytoplasm. This mRNA directs synthesis of Q but not the tri-peptide. How might E regulate synthesis of Q?

22-21　A particular mRNA molecule is detectable by a cloned c-DNA probe in liver cells but not in brain cells.
(a) How could you determine whether the gene is in the so-called "active" conformation in the liver cells?

(b) How could you tell whether the gene is in the nucleosomal conformation in liver cells?

22-22   A dihydrofolate reductase-deficient (*dfr⁻*) hamster cell line is transformed with a plasmid containing a *dfr⁺* structural gene, linked to the promoter of a hormone inducible gene, and another gene, *xgp*, which is a dominant selectable marker in this system. Transformed cells were selected for the ability to make Xgp, and all those cells selected also made Dfr. A transformed cell was treated with increasing doses of methotrexate, and a line of cells isolated which is resistant to 200-fold higher levels of methotrexate than the original transformant, and contains 200-fold more Dfr enzyme molecules.
(a) How would you determine how much of the increase in Dfr is due to gene amplification and how much to enhanced transcription of the gene? Why would you bother to look at the possibility of enhanced transcription, when you know that previous work with methotrexate-resistant mouse and hamster cells showed that the increased enzyme levels were proportional to (and apparently dependent solely on) the increased gene copy number?
(b) If you detected an increased copy number of a gene, how would you determine whether the amplified genes are integrated or extrachromosomal?
(c) Would you expect that your methotrexate-resistant cells have increased copies of the *xgp* gene as well as *dfr*, and why or why not? How would you test this possibility?

22-23   *Xenopus* oocytes were synchronously labeled at a late stage of oocyte development with ³H-uridine. Radioactive total cytoplasmic RNA isolated from the oocytes was hybridized with single-stranded filter bound total DNA from adult *Xenopus*. This RNA:DNA hybrid was competed with unlabeled total cytoplasmic RNA from gastrula-stage *Xenopus* embryos. The (hypothetical) result is shown in A. In a reciprocal experiment, radioactive gastrula-stage cytoplasmic RNA (gastrula-stage embryos which had been labeled with ³H-uridine) was hybridized to total DNA on filters. This RNA-DNA hybrid was competed with unlabeled late oocyte cytoplasmic RNA. The result is shown in B.
(a) What can you conclude about the RNA synthesized at these two stages of early *Xenopus* development.
(b) What information does this experiment give you about the role of transcriptional regulation in *Xenopus*?
(c) How would you attempt to learn more about the roles of

transcriptional regulation in early *Xenopus* development using similar techniques?

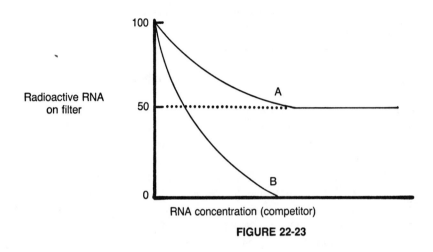

FIGURE 22-23

22-24   Several cases are known in which a single effector molecule regulates the synthesis of different proteins encoded in distinct mRNA molecules 1 and 2. Give a brief possible molecular explanation for each of the following observations made when the effector is absent.
(a) Neither nuclear nor cytoplasmic RNA can be found that hybridizes to either of the genes encoding molecules 1 and 2.

(b) Nuclear but not cytoplasmic RNA can be found that hybridizes to the genes encoding molecules 1 and 2.
(c) Both nuclear and cytoplasmic RNA but not polysome-associated RNA can be found that hybridizes to the genes encoding molecules 1 and 2.

22-25   A researcher wishes to study the expression of two kangaroo genes, *A* and *B*, in *E. coli*. In the experiment with gene A, she does a partial EcoRI digest of kangaroo DNA, inserts the pieces into the plasmid pBR322, transforms an *E. coli* deletion mutant lacking the *A* gene, and imposes a selection that requires that gene *A* is active. For gene *B* she uses reverse transcriptase to make DNA copies of the mRNA from kangaroo liver cells, which are known to produce large amounts of enzyme B. She then inserts these DNA copies into pBR322, transforms an *E. coli* mutant

lacking a functional B enzyme and selects for activity of B.
(a) What do you expect the results of these experiments to be and why? (Assume that in both cases the gene sequences have been inserted near a pBR322 promoter and in the proper phase).
(b) In a second experiment the kangaroo cells lacking function of the *A* gene were transformed with the plasmid containing gene *A* and kangaroo cells lacking a functional *B* gene were transformed with the plasmid containing gene *B*. The *A* gene but not the *B* gene is expressed. Suggest a possible explanation.

22-26   A gene family containing three genes is being studied. Messenger RNA molecules for the three genes appear in the cytoplasm at nearly the same time following turn-on of transcription. It has been hypothesized that a single transcript is made and that it is processed in three different ways (perhaps at random to yield the three mRNA molecules). The principal evidence is that the same sequence of four nucleotides is adjacent to the cap in all three mRNA molecules. The entire gene family has been cloned as a single unit in the plasmid pBR322. The plasmid DNA is partly denatured and all three mRNA molecules are added using conditions in which DNA-RNA hybrids form. Thus, some of the plasmid molecules have R loops, that is, bubbles in which one of the DNA strands is joined to mRNA rather than to its complement. The following molecules are observed in electron micrographs of these mRNA-containing plasmids (other forms are also seen). Could the hypothesis be correct?

FIGURE 22-26

ANSWERS

Chapter 1

1-1 (b).

1-2 The enzyme might be inactivated when the cells are broken; the concentration of the enzyme or of the reactants might be too low; the product might be unstable; the product might be destroyed by a second reaction; an enzyme inhibitor might be activated.

1-3 (a) Eukaryote.
(b) Only bacteria are prokaryotes.

1-4 Growth is for five generations. Thus, the initial number increases by a factor of 25 or 32 and the final concentration is $32 \times 10^5 = 3.2 \times 10^6$ cells/ml.

1-5 1, D; 2, A; 3, B; 4, C.

1-6 Log phase.

1-7 Exhaustion of nutrients; inability of atmospheric oxygen to dissolve in the growth medium as rapidly as it is being consumed; change of pH of the medium. The second item is the most important.

1-8 6.8×10^7.

1-9 Lysozyme—an enzyme that solubilizes the cell wall of most bacteria; lysis—breaking open a cell by any means.

1-10 Haploid—having only one copy of each chromosome; diploid—having two copies of each chromosome.

1-11 Primary cell—a cell that is either taken from an organism or is a descendant of one of these cells and is capable of growth for only a limited number of generations; established cell line—a population of cells derived from a primary culture but which has been altered, presumably genetically, so that it is able to multiply indefinitely.

1-12 (a) Genotype.
      (b) Pen-r.
      (c) No.
      (d) Pro⁻. Note that it is capitalized and roman.

1-13 Absolute defective—having a mutant phenotype in all conditions; conditional—having a mutant phenotype only in certain conditions; temperature-sensitive—having a mutant phenotype only above a critical temperature.

1-14 1.6×10^{-10}.

1-15 (a) No, because the defect might be in an enzyme required for glucose metabolism rather than for transport.
      (b) Yes, because after the lysozyme treatment, the cell must be permeable and allow glucose to enter the cell. The fact that, when the cell is made permeable, glucose metabolism occurs clearly shows that there is a permeability barrier at 42°C, which must be genetically determined.

1-16 In general, at least four. However, intragenic complementation may be possible, as described in the answer to problem 1-53.

1-17 (a) kyuQ.
      (b) A regulatory mutant that prevents synthesis of each gene product.

1-18 The enzyme is probably not the product of the gene.

1-19 That the G-gene and H-gene products probably join together to form an active macromolecular complex. The regions in which mutations occur (that can be compensated or are compensatory) are probably in the binding sites.

1-20 (a) *Leu⁻pro⁻*.

(b) Colonies on the Leu and Pro plates are Pro+ and Leu+ revertants, respectively. Thymine is not required since reversion for two markers would be unlikely. There are no colonies on the Thy plate for these would be double revertants, that is, Leu+Pro+.

1-21 1; For all markers, +; 2: *thr⁻*, others +, except for *pro*, which is unknown; 3: *arg⁻pro⁻*, others +; 4: *his⁻*, others +; 5: *thy⁻*, others +, except for *his* and *ser*, which are unknown; 6: *arg⁻*, others +.

1-22 All would be purple except for the triple mutant.

1-23 The original strain S had a *lac⁻* point mutation in the chromosome and contained a sex plasmid *F'lac*, which is temperature-sensitive in that it fails to replicate at 42°C. Variant 1 contains a Ts+ revertant of *F'lac*; 2 is a Lac+ recombinant; 3 has *F'lac* inserted into the gene for T1 sensitivity.

1-24 10^{-14} g = 6 x 10^{23} x 10^{-14} or 6 x 10^9 daltons. At 2 x 10^6 daltons/μm, the length of a strand of DNA is 6 x 10^9/(2 x 10^6) = 3000 μm. Considering the DNA as a cylinder, its volume is about 10^{-14} cm³. The volume of *E. coli* is about 2.4 x 10^{-12} cm³; thus, the ratio of DNA volume to cell volume is 0.004.

1-25 Yes. The colony contains about 7 x 10^8 bacteria. If every bacteria had divided every 25 minutes, there would have been about 10^9 cells.

1-26 Black produces only *B* gametes; gray makes both *B* and *b*, each with probability 1/2. Therefore, on the average, 1/2 of the offspring are *BB* and 1/2 are *Bb*. The probability of a single offspring being gray is 1/2. If there are two offspring, the probability of both being gray is 1/2 (for the first) x 1/2 (for the second) = 1/4. If there are three offspring, the probability of the first being gray and the others being black is 1/2 x 1/2 x 1/2 = 1/8. Since the gray could be the second or the third offspring, the probability of having only one gray is 1/8 + 1/8 + 1/8 = 3/8.

1-27 One. The gametes are *AB*, *Ab*, *aB*, and *ab*. These mate to form sixteen combinations, of which only *ab* x *ab* yields a homozygous recessive trait.

1-28   *RRtt* makes only *Rt* gametes. Since only *rr* is blue, there is no way (zero probability) to produce blue-eyed individuals.

1-29   Fat has the genotype *FF*; thin is *Ff*; *ff* is lethal.

1-30

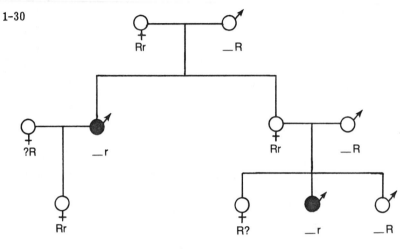

FIGURE A-1-30

1-31   Recessive, (a)(c). Dominant, (b)(d).

1-32   (a) The gene order is *a c b d*. (b) 0.15 x 0.04 = 0.006.

1-33   *b d a*.

1-34   The three classes of frequencies represent single, double, and triple crossovers. If *ABCd* and *abcD* are the products of a single crossover, the order must be *A B C D*.

1-35   (a) In a cross *AbDe* x *aBdE*, *AE* can result from a crossover anywhere between *A* and *E*. Since the map distance from *A* to *E* is 0.01 + 0.02 + 0.03 = 0.06, *AE* will occur at a frequency of 0.06. (b) To get *ABDE*, one needs crossovers in the intervals *A−B*, *B−D*, and *D−E*; the probability of getting all three crossovers is 0.01 x 0.02 x 0.03 = 0.000006. Therefore, the fraction of *AE* that will be *ABDE* is 0.000006/0.06 = 0.0001.

1-36   *e g f* is the gene order. The spacing between *e* and *g* is three times that between *f* and *g*.

1-37   (a) Red and long are both homozygous recessive; white and short are both heterozygous.
       (b) Homozygous dominant for either gene is lethal.

1-38   (a) Eye color.
       (b) The eye color and wing length genes are both carried on the X chromosome of the archaeopteryx.

1-39   (a) The genotype is his^- (Ts) trp^-leu^-.
       (b) The strain is his^-(Ts) and does not require histidine at 25°C.
       (c) The leucine requirement results from a deletion or double mutation.
       (d) The genotype is his^- (Ts) leu^-trp^+.
       (e) The genotype is $leu^-trp^-his^+$.
       (f) (1) *his, leu, trp* in that order. (2) Perform an interrupted mating. (3) Met^- is the counterselective marker, and prevents growth of the male bacteria.

1-40   The frequencies of linkage of donor genes are: *pur--pro*, 25/286 = 9 percent; *pur--his*, 159/286 = 56 percent; *pro--his*, 1/286 = < 1 percent. The gene order must be *pro pur his*, or *his pur pro*, depending on how you look at it. The *pur--his* distance must be much shorter than the *pur--pro* distance.

1-41   Statement 2.

1-42   Since *fda* is transferred early, transfer is counterclockwise. The map is 100 minutes long so it would take about 90 minutes.

1-43   *a*, 10; *b*, 15; *c*, 20; *d*, 30. Gene *d* is probably very near the *str* locus.

1-44   Between *c* and *d* because *x* is closely linked to both genes.

1-45   (a) True, because *a* enters before *c*.
       (b) False, because *b* enters before *c*.
       (c) True, because some crossovers will be between *a* and *b*.
       (d) False. See (c).
       (e) True, because *b* is between *a* and *c*.
       (f) True, because the a^- and b^- alleles are near, and on, the same DNA molecule.
       (g) True, because *b* enters after *a*.

1-46   (a) *c b d a.*

(b) Since *a* enters last, its entry limits the rate of appearance of recombinants. Thus the time-of-entry would be that on agar 2. Since a few recombinants would have the genotype a^+b^- and the agar allows only a^+b^+ cells to grow, the number of colonies would be slightly less than the values shown for agar 2.

1-47   With an Hfr, a portion of F is transferred first and the remainder is the last piece of DNA transferred.

1-48   *Pro* is a terminal marker. F is closely linked to *pro* (that is, all *pro⁺* recombinants are donors and hence F^+) and is therefore also at the terminus.

1-49   The *arg⁻* mutants have different map locations and must be in different genes; one is transferred before *met*; the other is transferred after *met*.

1-50   The gene order is *galA galB galC bio.*

1-51   (a) Six genes.

(b) Some of the closely linked mutations might lie in closely linked genes, that is, they might complement one another. Moreover, other genes in the pathway for which no mutants have yet been isolated might exist. Ten mutations distributed over seven or eight genes according to the Poisson distribution would stand a good chance of leaving one gene untouched.

1-52   Three genes. The groups are: (1,7), (2,4), and (3,6).

1-53   Generally, no. However, if the protein molecule is folded in such a way that it consists of two interacting regions (domains), a deleterious change in one domain, which destroys the interaction, could be compensated for by a change (which alone is deleterious) in the other domain. That is, the two incorrectly folded domains could interact correctly to yield a functional protein.

Chapter 2

2-1   (a, c, d) True. (b, e) False.

2-2   Nucleotide.

2-3   An exogenous nutrient is supplied to the organism from the environment; an endogenous nutrient is synthesized internally.

2-4   Carboxyl, hydroxyl, amino, sulfhydryl, disulfide.

2-5   An anabolic reaction consumes energy; a catabolic reaction generates energy.

2-6   (1, 3) anabolic; (2,4) catabolic.

2-7   Yes.

2-8   Either the sucrose cannot enter the bacterium or no intracellular enzyme is present that can cleave sucrose into its glucose units.

2-9   Algae. They can use atmospheric CO_2 as a carbon source.

2-10 The bacterium is able to use glycine as a carbon source in addition to its role as a protein precursor. However, glycine is not utilized as efficiently as glucose.

2-11 To generate ATP.

2-12 In ATP and other compounds containing high-energy phosphates, in glycogen, and in fats.

2-13 In the absence of O_2 the Krebs cycle is inactive, and less ATP is generated per mole of glucose during glycolysis. Also, oxidative phosphorylation, the major supplier of ATP, does not occur.

2-14 The environment of animal cells is constant and as long as the standard carbohydrate foods, which are mainly polymers of glucose, are available, there is no need to be able to metabolize a variety of sugars. Thus, in the course of evolution this ability has been lost or at least minimized.

2-15 Extracellular nucleases can break down nucleic acids (usually resulting from cellular decomposition) present in the environment. The products of nucleolytic action can then be utilized metabolically. The pancreas secretes nucleases into the intestine to break down nucleic acids in the digestive process.

Blood nucleases destroy nucleic acids released from viruses; this is essential since viral nucleic acid is frequently infective.

2-16 An enzyme is a protein; a coenzyme is a small organic molecule. Coenzymes usually carry out oxidation-reduction and phosphorylation reactions.

2-17 (a) Yes, as long as the radioactive atom is not one that would be removed in a J-to-G conversion.
(b) G and J are products of a branched pathway and the GeeB product catalyzes a step before the branch point.

2-18 The one in the methyl group.

2-19 A pH of 7 means a H^+ ion concentration of 10^{-7} M or 6×10^{13} ions/cm^3. Thus, there are 141 H^+ ions per bacterium. Statistical fluctuations probably do not produce much heterogeneity because they occur extremely rapidly and repeatedly. The internal pH is regulated by pH-sensitive pumps in the cell membrane.

2-20 Yes. If a complete reaction occurs anywhere, ΔG is negative.

2-21 Yes, because polymerization and depolymerization reactions are never the same; the latter requires intermediates (nucleoside triphosphates) which are not the products (nucleoside monophosphates) of depolymerization. In the case of protein synthesis and degradation, synthesis requires production of reaction intermediates (whose synthesis requires hydrolysis of ATP and GTP), which are not present in the enzymatic degradation of proteins.

2-22 They provide an alternate reaction pathway for which ΔG is large and negative. This is accomplished usually by synthesis of a reaction intermediate by a reaction for which ΔG is negative and whose conversion to the final product also has a negative value of ΔG.

2-23 The rate at which the sugar is broken down and the energy yield per mole (in particular, the number of ATP molecules per mole of sugar metabolized).

Chapter 3

3-1 Proline.

3-2 Those with charged groups other than the α-amino carboxyl group bind metal ions. An ionizable protein can bind a metal ion simply by forming an ionic bond. Many metal ions (such as Mg^{2+} and Cu^{2+}) form complexes with amino groups. The Hg^{2+} ion binds SH groups and is frequently bound to methionine and cysteine.

3-3 Two cysteines cross-linked by a disulfide bond.

3-4 Tryptophan, tyrosine, and phenylalanine.

3-5 All, using the CO and NH groups in the peptide bond. In addition, all polar amino acids can form hydrogen bonds via their side chains.

3-6 Three.

3-7 Cysteine: disulfide bonds; lysine: ionic bonds; isoleucine: hydrophobic bonds; glutamic acid: ionic bonds and hydrogen bonds. All could of course participate in van der Waals bonds.

3-8 There are only five peptide bonds because the C—N bond between serine and proline is not a peptide bond.

3-9 A carboxyl and an amino group.

3-10 The SH group of cysteine.

3-11 1', base; 3', OH; 5', phosphate.

3-12 Thymine, DNA; uracil, RNA.

3-13 A nucleoTide is a nucleoSide phosphate.

3-14 The 3' and 5' carbons.

3-15 (a) 3'-OH and 5'-P.
(b) One 3'-OH and one 5'-P group.

3-16 A random coil.

3-17 A hydrophobic interaction.

3-18 The bases are only weakly soluble; the little solubility that they have is conferred by the charged hydroxyl, amino, and keto groups carried, which can hydrogen bond with water.

3-19 (a) The compactness probably results from an ionic bond(s) between unlike charged groups.
(b) The molecule probably contains many groups having like charge and in the lower ionic strength of 0.01 M NaCl these groups repel one another.

3-20 It has sulfhydryl groups which are necessary for activity. If they are oxidized (mercaptoethanol prevents oxidation) or form disulfides, activity is lost.

3-21 If the salt concentration is high, the electric current is high and the temperature will rise until the gel melts.

3-22 No. To move an object x units north and then y units east brings the object to the same place reached by moving it first y units east and then x units north.

3-23 Electrophoretic mobility increases with charge but is reduced by friction between the moving molecule and the solvent molecules. Since valine has a longer side chain than alanine, it encounters greater friction and moves more slowly.

3-24 Shadowing and staining refer respectively to physical and chemical deposition of heavy atoms onto a macromolecule. In the negative contrast procedure transparent macromolecules are observed within a droplet of a solution of small molecules that absorb electrons strongly.

3-25 The crude measurement indicates that the molecular weight is between 168,000 and 228,000. Since 52,500 is an accurate measurement of the molecular weight of one subunit, there must be four subunits and the molecular weight must be 4 x 52,500 = 210,000.

3-26 (a) No, since differences in shape can affect mobility—that is, two proteins differing in both molecular weight and shape might have the same electrophoretic mobility.

(b) Yes, for all fragments have the same basic shape and same charge-to-mass ratio.

3-27   It is likely that the protein has subunits, though it is possible that the original sample consisted of two proteins having different molecular weights but precisely the same electrophoretic and chromatographic properties. This possibility is unlikely though. Assuming the protein has subunits, several possibilities exist—for example, one of each subunit with the larger one having twice the molecular weight of the smaller, equal numbers of each with the same ratio of the molecular weights, or three subunits with two copies of the smaller and one of the larger with the larger one having four times the molecular weight of the smaller one. The latter possibility could be distinguished from the first two since SDS-gel electrophoresis yields the molecular weights of the proteins. Gel chromatography will indicate if the initial sample contained molecules having two different molecular weights for these molecules would separate; if no separation is evident, the subunit explanation must be correct.

3-28   The relative distances migrated are roughly the ratio of log 1800/log 2600 with the smaller molecule moving faster. The ratio of the areas is (192 x 26,000)/64 x 1800) = 43.3 with the slower band having the larger area.

3-29   Vary the pore size of the gel so that the dependence of the mobility on the molecular weight would increase. Gel chromatography, which is sensitive only to molecular size, and ion-exchange chromatography, which separates proteins primarily by their charge, could also be tried.

3-30   The DNA is first coated with molecules of cytochrome c, which increases the effective thickness of the DNA. Shadowing at a low angle allows large amounts of metal to accumulate against the thickened molecule in the manner that snow accumulates against a fence in a storm, forming a drift or shadow. Depositing metal from directly above would cover everything.

3-31   In the negative contrast technique, the biological sample (which is transparent to electrons) is surrounded by an opaque material—in this case, tungsten atoms. The intensity of the image is inversely proportional to the thickness of the opaque material. Hence, a phage head filled with DNA appears as a solid white object against a darker background, but an empty phage head will become filled with phosphotungstic acid and will therefore appear as a light hexagonal outline against a dark background.

3-32   (a) If A is the shadow angle and l is the length of the shadow produced by a particle of height L, then $L = l \tan A$. The shadow angle can be accurately determined from the length of the shadow of a second particle of known dimensions.
(b) 492 angstroms. The fuzzy ones are probably partially collapsed molecules.

3-33   Such metals would adsorb more electrons than the sample, so contrast would be diminished.

3-34   The criterion should be that the fraction of all circles that are linked is independent of the DNA concentration in the sample prepared for electron microscopy. The use of very dilute samples minimizes overlap.

3-35   A flexible rod encounters less friction when moving through a fluid and thus has a higher sedimentation coefficient.

3-36   (a) The molecule is compact and probably is roughly spherical.
(b) The molecule is an extended rod.

3-37 $(1.7 + 1.3)/2 = 1.5$ g/cm^3.

3-38   To work this problem, you must know the average nitrogen content and carbon content of a protein, and you must assume that isotopic substitution does not change the volume of the protein (a good assumption). Proteins are about 14 percent nitrogen and 44 percent carbon. Therefore, 1.3 g of protein contains 0.182 g N and 0.572 g C. If the protein is labeled with ^{15}N, the N would weigh $(15/14)(0.182) = 0.195$ g and the volume of protein containing 1.3 g of ^{14}N-protein would weigh $1.3 - 0.182 + 0.195 = 1.313$ g, which is the density of ^{15}N-protein. ^{13}C-labeled-protein would have a density of 1.348 g/cm^3.

Chapter 4

4-1   (a) ACTAGTCCAGCTG.
       (b) TATATATATAT.

4-2   Bases: adenine, thymine, guanine, cytosine, uracil. Ribonucleosides: adenosine, guanosine, cytidine, uridine. Deoxyribonucleosides: deoxyadenosine, thymidine, deoxyguanosine, deoxycytidine. Ribonucleotides: adenylic acid, guanylic acid, cytidylic acid, uridylic acid. Deoxynucleotides: deoxyadenylic acid, thymidylic acid, deoxyguanylic acid, deoxycytidylic acid.

4-3   GC has 3.

4-4   (a) [purines] /[pyrimidines] = ([A] + [G]) = ([C] + [T]) = 1.
      (b) This ratio depends on the particular DNA molecule.

4-5   (a) The width is 2 nm and the length is 2×10^4 nm. Thus, the ratio is 10,000.
      (b) Such a molecule has a molecular weight of 40×10^6. Each base pair has a molecular weight of 672 so there are about 59,500 base pairs. Another way to calculate this is to note that the distance from one base pair to the next is 0.34 nm, so the number of base pairs is $2 \times 10^4/0.34 = 58,800$.

4-6   Since 1 μm corresponds to a molecular weight of 2×10^6, the length is 20 mm. The width is 2 nm, so the ratio is 10,000.

4-7   The mass/length of double-stranded DNA is approximately $2 \times 10^6/\mu m$; hence, the molecular weight is 32.8×10^6.

4-8   Each turn of the helix has about 10 base pairs or a molecular weight of 6700. Thus, there are $10^6/6700 = 149$ turns.

4-9   P-5'-AGAC-3'-OH; pApGpApC.

4-10 (a) The molecular weight of the bacterial DNA is 2 to 2.6×10^9 so there would be 2000 to 2600 genes.
      (b) There are about $8 \times 10^{10}/10^6 = 8 \times 10^4$ genes per average chromosome, so about 1.8×10^6 genes in total. Actually, only about 10 percent of mammalian DNA contains coding sequences, so the number of genes is more like 2×10^5.

4-11 The rule is that if $A_{260} = 1$, the DNA concentration is 50 μg/ml. Thus, the value of A_{260} is $32/50 = 0.64$.

4-12 The order should be I, II, III, since melting temperature increases with increasing (G+C) content.

4-13   Molecule I because of the long tract of GC pairs in II.

4-14   Molecule II, because the strands in I will preferentially form intrastrand hydrogen bonds.

4-15   Molecule II, since the long tract of GC pairs will require a higher temperature for dissociation of intrastrand hydrogen bonds formed after denaturation and cooling.

4-16   Since stacking forces are a major factor in the stability of DNA, and since the terminal bases have nothing with which to stack, the terminal base pairs are frequently unbonded. The extent of loose bonding increases with increasing temperature.

4-17   Note that there are 14 pairs of unpaired pairs in each molecule. Thus in a molecule having 100 base pairs, 14 percent will be unpaired. These are terminal bases because there is less of a tendency to stack at the termini because the terminal base pair can only stack on one side.

4-18   In A, because the melting temperature increases with the ratio of the solubility of deoxyribose to the solubility of the bases (a) and (b).

4-19   (a) and (b).

4-20   There are no counterions to shield the phosphates from one another. The strong electrostatic repulsion causes the strands to separate.

4-21   (b), because the effective solubility of the bases would be increased.

4-22   A double-stranded hairpin with AT pairs.

4-23   (a) If the ionic strength is very low, the sugar-phosphate backbone of the separated strands and all regions thereof are highly charged and mutually repulsive, thus preventing re-formation of base pairs. At high ionic strength both intra- and intermolecular base pairs form and therefore A_{260} decreases.
(b) When boiled A_{260} will increase to 1.37 but it will return to 1.00 at 25°C.
(c) For the linear molecule, after returning to 25°C some interstrand base pairs will form and A_{260} will decrease somewhat.

For the circular molecule, after returning to 25°C all base pairs will re-form and A_{260} will decrease to 1.00.

(d) The ratio will be 1. The bases in each strand form a palindrome; that is, the sequence from bases 1 to 7 is complementary to the sequence from 14 to 8. Therefore each strand will fold back on itself and form a double-stranded hairpin.

4-24   (a,b,c) The second substance is a better denaturant as it is more hydrophobic and destabilizes base stacking more effectively. (d) The second substance has more groups that can form hydrogen bonds with the bases and therefore is more effective.

4-25   (a) Nitrous acid cross-links the two strands.
(b) Some of the phosphodiester bonds are hydrolyzed at high temperature, so some single-stranded fragments do not spontaneously renature.

4-26   Strand separation is incomplete in 100 percent methanol.

4-27   Titration of amino groups prevents hydrogen bonding. Hydrophobic interactions maintain stacking and helicity but cannot cause base pairing.

4-28   Stacking keeps the bases in line so that adjacent bases can form hydrogen bonds with similarly oriented bases in a complementary strand.

4-29   In 7 M sodium perchlorate, the melting temperature must be lowered so much that at 25°C the DNA is partly denatured—in fact, it must be 48.9 percent hyperchromic.

4-30   Denatured DNA or some other ultraviolet-absorbing molecule is present in the sample. The observed value of T_m would be meaningful.

4-31   The absorbance increases would be greater for the native DNA. They could also be distinguished by passing a small amount through a nitrocellulose filter and comparing the value of A_{260} before and after filtration. Single-stranded DNA is removed from the solution.

4-32   By shielding the negatively charged phosphates from one another the positive ions of the salt stabilizes the DNA and thus raises the value of T_m.

4-33 No, unless there were fewer than about fifty base pairs.

4-34

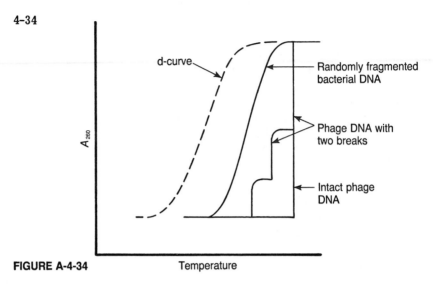

FIGURE A-4-34   Temperature

4-35 Hydroxyapatite chromatography; filtration through nitrocellulose filters; digestion with nucleases specific for single-stranded DNA.

4-36

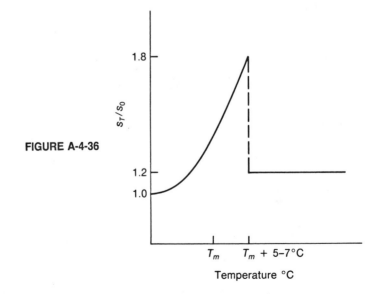

FIGURE A-4-36

4-37 The mixture is 45/50 HH and 5/50 LL before denaturation. After renaturation there will be $(0.9)^2 = 0.81$ HH, $(0.1)^2 = 0.01$ LL, and $2(0.9)(0.1) = 0.18$ HL, or 40.5, 9.0 and 0.5 μg of HH, HL, and LL, respectively.

4-38   Ten percent. (Note that if all sequences were common, only half would have a hybrid density).

4-39

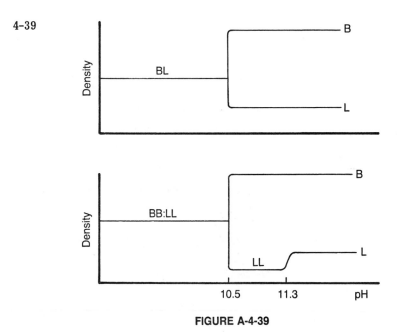

FIGURE A-4-39

4-40   (a) CpT, 0.15 (corresponding to ApG since both are written in the $5' \rightarrow 3'$ direction and the DNA strands are antiparallel); ApC, 0.03; TpC, 0.08; ApA, 0.10.
(b) If the strands were parallel, the known frequencies would be: TpC, 0.15; CpA, 0.03; CpT, 0.08; ApA, 0.10.

4-41

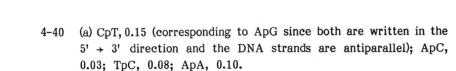

FIGURE A-4-41

4-42

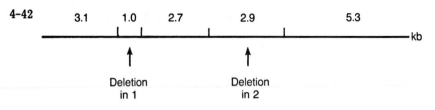

FIGURE A-4-42

4-43 The polyribopurine forms hydrogen bonds with the pyrimidine tracts. There will be two bands if the number of tracts is different in the two strands. The denser band contains DNA with more tracts and hence more bound polyribopurine. The separation increases as the molecular weight of the polyribopurine increases.

4-44 The strands of covalently closed circular double helical DNA cannot be separated by denaturation.
(a) The strands of DNA do not separate. Such DNA "renatures" (i.e. zips together again) rapidly. The DNA therefore bands at the density of helical, half-heavy, half-light DNA.
(b) The DNA bands at the density of helical, half-heavy, half-light DNA.
(c) The heavy and light strands of DNA now can separate and band separately as denatured DNA.

4-45 4 x 10 = 40 base pairs. Four nodes.

4-46 Ten base pairs are contained in one turn of a DNA helix. Thus, the molecule will be a supercoil having one node—that is, a figure-8.

4-47 Each turn of the helix contains ten base pairs, so the DNA acquires about five positive turns of supercoiling.

4-48 A linear DNA molecule can form a circle by cohesion of single-stranded termini; by recombination between redundant termini; and by exonucleolytic conversion of double-stranded termini to single-stranded termini, followed by hydrogen bonding between single strands.

4-49 Only twisting followed by joining.

4-50   Force needed is $F_{1/4}$, since a circle has half the length and twice the width of a linear molecule having the same molecular weight.

4-51   Dimers, trimers, and higher polymers—both circular and linear.

4-52   Less ethidium bromide can be bound by a supercoil than by a linear molecule. Therefore, the decrease in the density of a supercoil is less than that of a linear molecule.

4-53   No, because ethidium bromide can intercalate between any two base pairs.

4-54   (a) 1/2 (1.592 + 1.556) = 1.574.
        (b) 1/3 (1.592 + 1.556 + 1.556) = 1.568.

4-55   The supercoiled form has many unpaired bases (that is, single-stranded regions) which result from the unwinding of the helix.

4-56   Yes, if the density of the substance bound differs from the density of the DNA. If the substance binds preferentially to single-stranded DNA, more will bind to the supercoil and separation will occur.

4-57   The denser band is clearly a supercoil. When there is an average of one break per molecule, the fraction unbroken is $1/e$ or 37 percent. Therefore, the 2:1 ratio would become $2(0.37):[1 + 2(1 - 0.37)]$, or 0.74:2.26, or approximately 1:3.

4-58   Naturally occurring DNA is negatively supercoiled. Ethidium bromide adds positive supercoiling. Thus, the compact supercoil progressively loses its supercoiling until a slower moving nontwisted circle has formed. As more ethidium bromide is bound, the DNA becomes progressively positively supercoiled and hence more compact so the value of s increases.

4-59   (a) Protein molecules bind to nitrocellulose and double-stranded DNA does not. Therefore, a DNA molecule to which protein molecules are bound will adhere to the filter by means of the protein molecules.
        (b) The amount of DNA bound when the repressor is bound must be measured.

4-60   Protein A may bind very weakly to superhelical DNA and may possibly introduce superhelical twists of the same sense as those in the DNA molecule. Alternatively, it binds strongly but does not alter *s* other than by increasing the molecular weight. Protein C may bind weakly, and thereby remove some superhelical twists, but it may also be a highly asymmetric molecule that increases the frictional coeffficient of the DNA molecule decreasing *s* (by cancelling the effect of superhelicity). Protein B introduces superhelical twists in the opposite sense from those already in the DNA.

4-61

FIGURE A-4-61

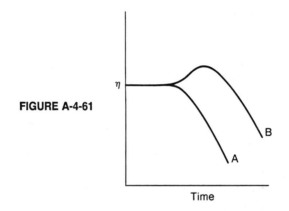

4-62   The irradiation probably damages the base so that the N-glycosylic bond (attaching the base to the deoxyribose) is broken by alkali. Once this bond is broken, the phosphodiester bond can also be broken by alkali. Thus the site of damage at each base is converted to a single-strand break. The slowly moving material consists of the DNA fragments formed by this breakage.

4-63   First denature the DNA and add an exonuclease. Alternatively, digest extensively with an endonuclease and then add an exonuclease to complete the hydrolysis. Spleen and venom phosphodiesterases are useful.

4-64   The enzyme fails to bind to the DNA at such high ionic strength.

4-65   The enzyme is inhibited by the nucleotides produced in the reaction.

4-66   The DNA molecule must contain something other than

single-stranded DNA one-third of the way from the 5'-P terminus. Possibilities include protein or RNA linkages within the polynucleotide chain, a protein molecule bound to a sequence of DNA bases, a short sequence of double-stranded DNA or RNA, or a hairpin-like structure (a palindrome, or inverted repetition). The boiling experiment indicates reversible thermal disruption at high but not low ionic strength; this strongly implicates the hairpin-like structure.

4-67   (a) Venom—none; spleen—Ap, CpGp, Up; T1—ApCpGp, CpUpUpCp.
       (b) Venom—A, pU, pC; spleen—Ap, UpC; pancreatic—Ap, UpC.

4-68   (a) A few single-stranded regions are produced at temperatures before which an increase in optical density is detectable.
       (b) Random breakage and re-formation of hydrogen-bonded regions occur continually. Because of the possibility of free rotation of the unbroken strand opposite a single-strand break, the fraction of time that the bases of any region are unpaired will increase in the vicinity of a single-strand break, thereby producing a single-stranded region. This will be more effective as temperature increases.
       (c) Supercoiled DNA molecules possess single-stranded regions. A single-strand break will first be made and then will become a double-strand break according to the reasoning in part (b).
       (d) A mismatched base pair generates a single-stranded region. Half of the renatured molecules will contain a mismatched base pair. Thus, only the hybrid DNA molecules are attacked by S1 nuclease.

4-69   Possibilities include a circular DNA molecule, altered termini (for example, altered 3'-P or 5'-OH groups), a blocked terminus (for example, blocked by an ester), and a hairpin producing a double-stranded terminus.

4-70   It is probably an exonuclease acting on only one type of terminus (3' or 5'). Thus, single stranded DNA is produced at both ends. Both the single-stranded regions and the mononucleotides have a higher value of A_{260} than double-stranded DNA, which accounts for the increase in A_{260}. The single-stranded ends are flexible and compensate for the decrease in molecular weight of the entire molecule in determining the s value.

4-71   Treatment with phosphatase and polynucleotide kinase results in ^{32}P-labeling of the 5'-P termini. Annealing plus ligation produces a covalent circle. The ^{32}P is now attached to the base at the 3'-OH terminus. Enzyme digestion results in cleavage of the 3'-terminal base, now labeled with ^{32}P. Therefore, guanine is at the 3'-OH terminus and, by complementarity, cytosine is the 13th base from the 5'-P terminus.

4-72   Adenovirus DNA carries an inverted terminal repetition; that is, the sequence of bases along each strand is 3' ABCD...D'C'B'A'-5', where letters followed by primes denote the bases which can form base pairs with non-prime bases.

4-73   Physical experiment: Denature, dilute, and subject to renaturing conditions. The longer strand of molecule II would form a circle. Then centrifuge. Molecule II would yield two types of molecules—slowly moving, linear, single-strand and faster moving, circular single strand. Molecule I would yield only a single species.
       Enzymatic experiment: Use radioactive DNA and treat with a 5'-P-specific and a 3'-OH-specific exonuclease active only against single-stranded DNA. For molecule I, only the 5'-specific enzyme can release mononucleotides; for molecule II, both enzymes can do so.

4-74   The free 2'-OH groups allows the β-elimination reaction to occur.

4-75   Single-stranded DNA can be distinguished from single-stranded RNA by enzymatic digestion by specific nucleases; RNA is hydrolyzed by alkali but DNA is not; the diphenylamine test distinguishes ribose from deoxyribose; acid hydrolysis followed by chromatography or electrophoresis distinguishes uracil from thymine.

4-76   (a) The 5'-fragment is unique in that it will yield a diphosphoryl nucleotide (pXp) upon exhaustive degradation with alkali.
       (b) Only that the 3'-terminal fragment contains 19 bases.
       (c) Only that methyladenosine is on the 5' end of the fragment.
       (d) Fragment 1 is 3'-terminal because it has adenosine as a base hydrolysis product.
       (e) Fragment 3 is 5'-terminal because it contains the methyl-A, which is known from (c) to be 5'-terminal.
       (f) Fragment 1: 5'-A; 3'-A. Fragment 2: 5'-C; 3'-G. Fragment 3: 5'-methyl-A; 3'-G.
       (g) CpUpApGp.
       (h) Current information is

5'-[methyl-A(2U,2C)Gp][CpUpApGp][A(2C,1A)pCpCpA]-3'

     Fragment 3   Fragment 2   Fragment 1

The parentheses indicate that the composition but not the sequence of the segment is known. Assume a 3'-pCpCpA sequence with two G's distributed between fragments 3 and 2 or fragments 2 and 1. The best approach to finding the two missing G's is to digest the large (19-base) fragment with pancreatic RNase and determine whether two GpGp's or one GpGpGp is generated, and identify other nucleotides that might be linked to it (or to them). The sequence of fragments 1 and 3 can be obtained by defining the products released by digestion with pancreatic RNase or with phosphatase and venom phosphodiesterase.

4-77   (a) Equimolar amounts of Ap, Cp, Gp, and Up indicate a tetranucleotide, since one Gp is the maximum possible after treatment with T1 RNase. Therefore, the sequence at this stage is (A,C,U)Gp, in which the sequence of A, C, and U is not known.

(b) C must be 5'-terminal. Therefore, the sequence is either CpApUpGp or CpUpApGp.

(c) If the sequence were CpUpApGp, pancreative RNase would yield free Cp and Up which is not the case. Thus, the sequence is CpApUpTpGp. The reaction scheme is shown in Figure A4-77.

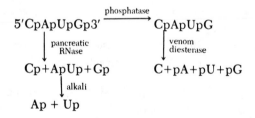

FIGURE A-4-77

4-78   (a) The 3' end.

(b) 5'-UpCpCpA-3'.

4-79   (a) Bases far from the 5' end could not be identified because each strand would be cleaved more than one time.

(b) The sequence of the DNA in panel (i) is TCGAAGCCTTAAC. That of panel (ii) is GACCGGAATTGCA.

(c) The short run gives the sequence GGTACTAGGGTATCAAT. The long run gives the sequence CAATGGATCGTCAGATC. Presumably the

CAAT is common to both so the sequence is GGTACTAGGGT-ATCAATGGATCGTCAGATC.

4-80 (a) 5'-AGCTTATGCGC_TGGGGCAA-3'.
      (b) 5'-AGCTTTGCCC_AGGCGGATA-3'.
      (c) The sequence known from the cleavage site, namely,
      5' AGCTT...3'.
      (d) The sequence is

       5'A G C T T A T G C G C C̲ T G G G G C A A 3'
             3'A T A C G C G G A C̲ C C C G T T T G C A 5'

The two underlined C's are missing from the ladder and can only be inferred from the presence of the G's.

(e) These C's are probably modified in a way that interferes with the chemical reactions used in the procedure. A good guess would be methylation, as 5-methylcytosine appears frequently in DNA. In fact, it is known that this base does not to respond to the hydrazine-cleavage procedure.

4-81 Results 1, 4, and 5 show that there is some single-stranded DNA. Result 2 shows that there is some double-stranded DNA; the shift of 0.010 g/cm^3 rather than 0.014 g/cm^3, suggests that about 70 percent of the DNA is double-stranded. Result 3 suggests that the molecule is circular; this is confirmed by result 6. Result 7 shows that the two component strands do not have the same length. The molecule consists of one circular strand to which is hydrogen-bonded a linear strand whose molecular weight is roughly 70 percent that of its circular strand.

4-82 Result 1 indicates that the DNA is double-stranded. Result 7 shows there is an interruption in one strand. Result 2 shows that the interruption is not in the center. Result 3 indicates, by virtue of circle formation following enzymatic treatment, that the DNA is linear and terminally redundant. Result 4 shows that the interruption is a nick rather than a gap and that it has 5'-P and 3'-OH termini. The 5'-P group at the nick is confirmed by result 8. The linearity of A is shown by result 9. A is a terminally redundant linear molecule with a nick in one strand, not in the center. B is a nicked circle formed by joining of single-stranded termini and having an additional nick in one strand. C is a nonsupercoiled covalent circle.

4-83   The DNA is double-stranded without single-stranded termini but
        with a single-strand break in each strand. These strand breaks
        are each located 1/3 of the distance from the 5' end. The strand
        binding poly(G) has guanine at the 5' terminus; the other strand
        has adenine.

4-84   (a) Result 1 suggests that RNA is present; result 2 shows that
        the RNA is not simply a single strand. Result 3 suggests a
        supercoil or single-stranded circle; result 4 eliminates the
        latter. Results 4 and 5 show that the RNA is probably contained
        in one strand; results 6 and 8 show that the DNA strand is
        circular. Result 7 shows that equal amounts of RNA and DNA are
        present. The molecule is probably a supercoil with one DNA strand
        and one RNA strand. A is a circle with the DNA strand nicked. B
        is a circular DNA strand (C) plus a linear RNA strand. D consists
        of monoribonucleotides.

Chapter 5

5-1   (a) The polar amino acids are arginine, asparagine, aspartic
        acid, cysteine, glutamic acid, glutamine, histidine, lysine,
        serine, threonine, and tyrosine. The nonpolar amino acids are
        alanine, glycine, isoleucine, leucine, methionine, phenylalanine,
        proline, tryptophan, and valine.
        (b) Isoleucine is more nonpolar than alanine, because it has a
        long, nonpolar side chain. Proline is unable to form a proper
        peptide because it lacks a free amino group.

5-2 $32,000/110 = 299.$

5-3   The peptide bond.

5-4   Polyalanine is uncharged so it will stay in the α-helical
        configuration. Polyaspartic acid will be an a helix only below pH
        3.8 when the carboxyl group of the side chain is not ionized.

5-5   Cysteine—disulfide; arginine—ionic; valine—hydrophobic;
        aspartic acid—ionic and hydrogen bonds. All of course could be
        in van der Waals bonds.

5-6   Those with nonpolar side chains (see answer 5-1) tend to be
        internal. Those with polar side chains tend to be on the surface.

There is little preference for glycine, which has no side chain, for the nonpolar rings with a very weak charge (e.g., tryptophan) and for cysteine. Two cysteines can form a covalent bond called a disulfide bond; when this bond forms, the two cysteines are they are internal.

5-7 (a), (c), and (f) are true. The remainder are false.

5-8 Think about the amino acids that would tend to form ionic bonds and hydrophobic bonds. The ionic bonds in (a) and (b) will be pH-sensitive. The hydrophobic bonds will not. (b) will have an internal hydrophobic cluster of isoleucine and two phenylalanines. (c) will stack with the rings parallel. (d) will be extended when the side chains are charged and an α helix when they are not charged; thus, this structure will also be pH-sensitive.

5-9 Both will tend to be near an amino acid of opposite charge.

5-10 Formaldehyde reacts with amino groups, thereby preventing hydrogen-bonding.

5-11 (a) Since each molecule of protein X must contain at least one residue of the rarest amino acid, the number of amino acids in X cannot be less than $100/0.5 = 200$.
(b) The sedimentation rate is an approximate measure of the molecular weight, which, in turn, is an approximate measure of the total number of amino acids in a protein. A sedimentation rate slightly less than that of the particular protein implies a number of amino acids less than 267. Evidently, protein X does contain 200 amino acids rather than some integral multiple of 200, and the number of methionines per molecule must be simply 3.

5-12 Once the enzyme has cleaved a terminal amino acid, it may then either cleave the next amino acid or remove the terminal amino acid from a second protein. Hence, when two amino acids per protein, on the average, have been removed, there are proteins which have had one, two, and three amino acids removed. Since only alanine has been found, the terminal tripeptide must consist of three alanines.

5-13 Set 3, in a hydrophobic cluster.

5-14   The α helix and β structure, because in these structures the peptide groups are hydrogen-bonded to one another.

5-15   (a) An extended coil but not an α helix, because arginine is charged.
        (b) In strong alkali (pH > 13) it is uncharged and the α helix should predominate.

5-16   (a) 4 and 5, 9 and 10, 11 and 12.
        (b) Three.

5-17   Val Ala Phe Leu Lys Met Trp Pro Arg Val Met Gly.

5-18   (a) Yes. A van der Waals attraction would favor aggregation. This would be aided by a hydrophobic interaction if the side chains on the surface were very nonpolar.
        (b) No. For geometric reasons the hydrophobic regions probably cannot come into contact.
        (c) No. Charge repulsion will effectively counteract any tendency to form a hydrophobic cluster.
        (d) Yes. If the geometry is appropriate, the alternation of unlike charges will allow unlike charges in the two molecules to attract.

5-19   The C-terminal amino acid is valine. Once this is removed, the enzyme can remove the adjacent amino acid.

5-20   Only hydrogen bonds.

5-21   Structural proteins.

5-22   About 75 nm.

5-23   (b), (c), (d), (e), and (f) are true. Note that (e) is true only if the substances have a common chemical feature.

5-24   No. In general, enzymes must be slightly flexible in order to adapt their shape to the substrate and to carry out the required chemical reaction.

5-25   The flexing of an enzyme about a substrate.

5-26   Feedback inhibition, enzyme inhibitors, enzyme activation, zymogen activation.

5-27 Charged groups, normally neutralized by ions present in the solvent, interact and thereby change the shape of the protein from an active to an inactive form.

5-28 Foaming increases the surface-to-volume ratio of the solution. This increases the probability that an enzyme molecule will be at the surface and subjected to disruption by the force of surface tension. It is worthwhile to determine for yourself how these forces can destroy activity. If the enzyme contains essential sulfhydryl groups in or near the active site, inactivation may result from the greater rate of oxidation of these groups in the highly aerated foam.

5-29 Hydrophobic interactions are the most common. All types of bonds can do it though.

5-30 The enzyme has three subunits if, in urea, the molecular weight drops threefold. When active and inactive forms are mixed, dissociated, and reassociated, the subunits will reassociate at random to produce molecules having some active and some inactive subunits; those having 0, 1, 2, and 3 active subunits arise at frequencies 1/8, 3/8, 3/8, and 1/8, respectively. If the activities are 12.5 percent, 50 percent, and 87.5 percent, there must be 3, 2, and 1 active subunits, respectively, if the molecule is to be active.

5-31 The enzyme could be a multisubunit protein in which one defective subunit is sufficient to eliminate enzymatic activity.

5-32 (a) Gel chromatography or gel electrophoresis.
(b) Add proteins of known molecular weight and obtain a standard curve in which either the volume of the effluent (for chromatography) or the distance migrated (for electrophoresis) is plotted against molecular weight.

5-33 Activity can be regulated by dissociation and association of subunits. For example, in a protein solution at low concentration, an inactive monomer will be the major species. Also, less genetic information is required to produce the complete protein.

5-34 Enzymes whose activity is controlled by the presence of small molecules (for example, allosteric enzymes) must contain several binding sites and therefore tend to be large. The polymerases that carry out several functions (for example, polymerase, 5' → 3' exonuclease, and 3' → 5' exonuclease) must have numerous active sites. A typical protein with a single active site usually has a molecular weight of 20,000 to 60,000. With active sites for binding of bases in polymerization, plus the other activities, a molecular weight in the 100,000 to 200,000 range is expected.

5-35   Metal ions or small molecules are part of the active enzyme; usually these are the electron acceptors.

5-36   (a) The charge on one or more amino acids in the active site is changed.
(b) At pH 11 the protein denatures. There is an amino acid in the active site whose charge changes between pH 5 and 7.

5-37   If the binding of the ion is not too tight, a high concentration of the ion is needed to have bound ions. The ion exchanger will remove these ions. In gel chromatography the ions will move more slowly than the protein and will be separated. The solution to the problem is to try gel chromatography with a column that is equilibrated with the ion and to elute with a solution of the same ion concentration.

5-38   Above pH 4 the polyglutamic acid is charged and the α helix is disrupted; this accounts for the structural change. The viscosity decreases because at the higher salt concentration in A the charges are shielded and the molecule collapses. At the low salt concentration in B the charges repel one another and the polypeptide is more extended than when it is in the helical configuration.

Chapter 6

6-1   The collagen of the young animals has fewer cross links and can be dissociated by shielding of the ionic bonds by a high salt concentration.

6-2   The positively and negatively charged amino acids produce the quarter-stagger array which can become indefinitely long.

6-3   Without the RNA to fix the length, the reconstituted particles, which have a slightly different structure, grow indefinitely long.

6-4 There are no cross links, because a nitrile in the sweet pea seed inhibits the enzyme that forms these bonds.

6-5 Collagen is a major constituent of the walls of blood vessels. With hydroxyproline deficiency the collagen is defective and localized rupture of blood vessels occurs.

6-6 To render the cell wall somewhat resistant to digestive enzymes in the environment.

6-7 Bacteria are in contact with many noxious environmental substances which can damage a variety of biological molecules. A multiple layer that consists of different components ensures that if one layer is damaged, the physical integrity of the cell need not be lost.

6-8 It probably enables the DNA molecule, whose length is about 400 times that of a bacterium, to fit neatly inside the cell. It may also be important in guiding daughter DNA molecules to daughter cells.

6-9 Chromatin—the DNA-histone complex present in eukaryotic chromosomes; histones—a class of five proteins bound to DNA in chromatin; nucleosome—a DNA-histone complex consisting of a DNA-histone core particle, linker DNA, and histone H1; core particle—a DNA-histone complex consisting of an octamer (two copies each of H2A, H2B, H3, and H4) and a 140-base-pair DNA molecule; linker DNA—the segment of DNA that is not bound to a histone octamer.

6-10 Eukaryotic DNA is enormous and there are many different DNA molecules in each nucleus. The existence of chromosomes prevents tangling during cell division and nucleosomes are the first step in reducing the effective length of each DNA molecule. It has recently been suggested that the nucleosome structure may play a role in regulating gene expression.

6-11 Lys-Trp-Lys and Lys-Tyr-Lys are very effective. The symmetric lysines bind to the phosphates and the tryptophan and tyrosine stabilize binding by stacking to adjacent pyrimidines.

Lys–Gly–Lys works but weakly because there is no stacking stabilization. Lys–Trp–Gly will bind to the DNA only very weakly, if at all, because of the lack of symmetry and Lys–Gly–Gly will have no affinity for DNA.

6-12 (a) None. Results (1) and (2) test only one type of linkage. Result (3) is independent of the nature of the linkage. (b) Complete digestion with a protease and a nuclease and identification of a component which is neither a free amino acid or a free nucleotide but consists of one amino acid and one nucleotide. One could show that the amino acid and nucleotide are joined by observing co-migration during electrophoresis or chromatography; since amino acids and nucleotides do not bind noncovalently, one can assume that the amino acid and the nucleotide are covalently joined in this complex.

6-13 The lack of sequence specificity suggests that either the phosphates or deoxyribose are involved in binding. The effect of 1 M NaCl indicates ionic binding. Thus, the phosphates are likely to be very important components of the binding site.

6-14 The structure of the protein might vary with salt concentration and have a better fit to the DNA in 1 M NaCl. Alternatively, it may bind as a dimer and the dimer form is favored over the monomer in 1 M NaCl. More likely, the protein has a negatively charged amino acid on its surface and hence is repelled by the negatively charged phosphates in the DNA. In 1 M NaCl these charges would be shielded from one another.

6-15 These molecules are intercalating agents and would move the DNA bases apart. Since the Cro protein binds stereospecifically, it is likely that the Cro–DNA complex would not form. Also, the intercalating agent might block specific contact points.

6-16 It is very compact. Removal of the RNA and protein needed to maintain the compact structure will reduce s and increase the viscosity.

6-17 When there is a single binding site, it is probably located at or near the junction of all of the subunits. This location is certainly unlikely if there are several identical binding sites. If the number of binding sites equals the number of subunits, one might guess that the binding sites are far from all regions of contact between the subunits. If the number of binding sites is half the number of subunits, each binding site probably includes the contact region of two subunits.

6-18   Antibodies have the function of inactivating or removing unwanted molecules or particles. One way antibodies accomplish this is by selective precipitation. Since many antigens have two binding sites, an antibody with two identical binding sites is able to form a mixed molecular network that precipitates. The precipitate is then destroyed by special phagocytic cells that ingest and digest the precipitate.

Chapter 7

7-1   (a,c,e), true; (b,d), false. (The advanced student familiar with the phenomenon of protein and mRNA processing should recognize that (a) is not always true.

7-2   As a simple repeating polymer consisting of a single unit, it cannot carry information for an amino acid sequence.

7-3   (a).

7-4   (a) The solution containing 10^{-7} mg/ml DNA would contain 0.02 percent of 10^{-7} mg = 2×10^{-14} g protein. The molecular weight of a protein containing 300 amino acids is 300 times the "average" weight of an amino acid, or approximately $300 \times 100 = 3 \times 10^4$. We can set up the proportions

$$\frac{X \text{ molecules}}{2 \times 10^{-14} g} = \frac{6 \times 10^{23} \text{ molecules per mole}}{3 \times 10^4 \text{ g per mole}} \quad ,$$

or $X = 4 \times 10^5$ molecules.
(b) The numbers do not exclude the possibility that the transformation is protein-mediated, since the maximum estimate of the number of protein molecules(4×10^5) exceeds the number of transformants 400-fold.

7-5   Repeat the transformation with other genetic markers.

7-6   Isolate DNA and centrifuge it in a sucrose gradient. Fractionate the centrifuge tube and test each sample for transformability using a very low DNA concentration. Try to find two genetic markers that co-transform—that is, two for which the recipient bacterium can acquire both markers even when the DNA is very dilute. You will observe that the genetic element that carries linked markers always sediments more rapidly than the bulk of the elements that carry only one marker. If this can be shown for several pairs of markers, you will have shown that transformability sediments with DNA and that co-transformability requires DNA of higher molecular weight. This should be pretty convincing. You could then also break the DNA molecules (perhaps by hydrodynamic shear) and show that as the average molecular weight of the DNA decreases, there is no loss of transformability of individual markers but that linkage is lost.

7-7 Show that transformability as a function of pH follows curves for DNA denaturation rather than protein denaturation. Show that transformation occurs at an ionic strength (e.g., 0.3 M) at which DNA-protein interactions are very weak. Separate protein and DNA by gel chromatography.

7-8 Isolate the DNA and centrifuge to equilibrium in buoyant CsCl. Since DNA and protein have different densities they will come to equilibrium at different positions in the gradient. Fractionate the centrifuge tube. The fractions containing DNA only or protein only can be tested separately for transforming activity.

7-9 (a) Some of the protein is injected into the bacterium.
(b) Some phages adsorb but fail to inject their DNA. Presumably they are defective.
(c) The ^{35}S-containing parts of the phage are removed from the bacterium, and phage production is still possible. Thus they are no longer needed for phage production.

7-10 Blending has apparently broken open the cells and the fragments of the cell wall plus any material that has adhered to them are in the pellet. Thus you could show that protein is in the pellet (plus some uninjected DNA) and that injected DNA is in the supernatant.

7-11 (a) (1) would not, because with this phage parental genetic material might rarely be packaged. (2) is not very convincing, because the amount of ^{32}P is so small. (3) is relatively convincing.
(b) Yes, if parental DNA is never packaged.

7-12   Parental DNA is used as a template for synthesis of progeny DNA and is rarely packaged into a phage head.

Chapter 8

8-1   (a) and (e) are true.

8-2   (3).

8-3   Deoxynucleoside triphosphates. Addition to a 3'-OH group; requirement for a template.

8-4   Primer—a nucleotide bound to DNA and having a 3'-OH group; template—a polynucleotide strand whose base sequence can be copied.

8-5   RNA serves as a primer.

8-6   (a) 5'-P.
        (b) 3'-OH.
        (c) 5'-P.

8-7   No initiation—$dnaA^-$ and $dnaC^-$; no synthesis—$dnaB^-$, $dnaE^-$, and $dnaG^-$; no initiation of precursor fragments—$dnaG^-$; reduced joining of precursor and uracil fragments—lig^-, $polA^-$; no DNA synthesis because of lack of deoxynucleotides—$dnaF^-$; no effect—$polB^-$.

8-8   (a)   **TABLE A-8-8**

	Fraction of DNA that is		
Generation	$^{15}N^{15}N$	$^{15}N^{14}N$	$^{14}N^{14}N$
0	1	0	0
$\frac{1}{2}$	$\frac{1}{3}$	$\frac{2}{3}$	0
1	0	1	0
$1\frac{1}{2}$	0	$\frac{2}{3}$	$\frac{1}{3}$

(b) If the DNA were not broken, the density would shift continuously; at one generation the DNA would have the density of $^{15}N^{14}N$.

(c) Same as (b) as long as the daughter helix remained attached to the parental DNA; at one generation, two bands—HH and LL—would be seen.

8-9 A second round of replication begins before the first is completed.

8-10 X prevents initiation of a round of replication but has no effect on movement of the replication fork. Rifampicin or chloramphenicol are examples.

8-11 Statement 1.

8-12 Certainly the molecules having intermediate density exist. However, a problem exists in detecting these molecules. Since the chromosome is broken at random, the fraction of the DNA in the fork that is hybrid DNA can range from nearly zero to nearly one. Thus, the density of molecules containing the replication fork will range from heavy to hybrid. The molecules will be distributed uniformly in the intermediate density region rather than in a discrete peak. Furthermore, these molecules comprise so little of the material that they are not normally detected.

8-13 Polymerase III is responsible for the addition of all nucleotides in the growing fork; polymerase I removes RNA from the 5' end of precursor fragments and replaces it with deoxynucleotides.

8-14 Priming of precursor fragments occurs in the presence of rifampicin, which inhibits RNA polymerase and does not occur in a mutant with a defective dnaG gene, which encodes primase.

8-15 The 5' → 3' activity removes ribonucleotides from the 5' termini of precursor fragments and the 3' → 5' exonuclease removes a base that has been incorrectly added to the growing end of a DNA strand.

8-16 Almost all nucleases require the Mg^{2+} ion for enzymatic activity.

8-17 The ligases form an active intermediate containing AMP by reacting with either NAD or ATP.

8-18 A collection of oligonucleotides of varied composition derived from degradation of DNA and RNA during purification. These oligonucleotides served as primers.

8-19　RNA polymerase is used only to initiate a round of replication. Since replication is bidirectional, there are two events per round of replication that utilize RNA polymerase. Primase initiates each precursor fragment. The molecular weight of one *E. coli* DNA molecule is 2.7×10^9. Only one strand equivalent is replicated by lagging strand synthesis or a molecular weight of 1.35×10^9. A precursor fragment has an average molecular weight of 10^6, so roughly 1350 events are initiated by primase.

8-20　(a) Breathing.
(b) Without a helicase, replication depends on breathing as the sole cause of unwinding. Thus, replication is much slower when there is no helicase. There is no effect on fidelity.

8-21　A helicase unwinds a helix. The ssb proteins prevent the helix from rewinding (forming intermolecular hydrogen bonds) and prevent intramolecular base pairing from occurring.

8-22　*In vitro* the concentration of pyrophosphate produced by polymerization is too low to cause the equilibrium to shift toward hydrolysis. *In vivo* there are many intracellular reactions in which ATP is cleaved, so that the pyrophosphate concentration is much higher than in an *in vitro* reaction.

8-23　Nick the DNA lightly with an endonuclease and carry out nick translation. Note that you cannot start copying unnicked DNA from one end because there is no primer; you cannot even add a primer because in general you would not know the terminal base sequence.

8-24　(a) *Thy⁺* bacteria make dTMP from dUMP, not from thymine; *thy⁻* bacteria start with thymine.
(b) Since one must add uracil or uridine, thymine cannot be used. Thymidine would work though.

8-25　(a) The 5' end.
(b) The 3'-terminal ribonucleotide is A.
(c) Its nearest-neighbor deoxyribonucleotide is mainly dG, sometimes dA, and rarely dT (but dT might not be significantly above background error level). Thus, two is the minimum number.

8-26   The molecular weight of *E. coli* DNA is 2.6×10^9, so there are 8.6×10^6 nucleotides. Complete synthesis occurs in $40 \times 60 = 2400$ seconds. Two replication forks are used so the required time for complete synthesis is 4800 seconds/fork. Thus, the polymerization rate is $8.6 \times 10^6 / 4.8 \times 10^3 = 3600$ nucleotides/sec.

8-27   Assume bidirectional replication and that both replication forks move at the same rate. Then the order is determined by the distance of each gene from the origin. Thus, the order is *AGBFCDE*.

8-28   The two strands of DNA are antiparallel and the DNA polymers can only add a nucleoside 5'-triphosphate to a 3'-OH terminus. The problem is solved by the cell by synthesizing short precursor fragments in a direction opposite that of movement of the fork.

8-29   Pulse-labeled DNA appears in small fragments when sedimented in alkali. If pulsed and chased, the radioactive DNA has the same average size as the major portion of the cellular DNA.

8-30   Measurement of the fraction of pulse-labeled material that is in small fragments. Hybridization does not give information because of bidirectional replication.

8-31   Sequence 1.

8-32   Gap-filling by polymerase I; then, joining by ligase.

8-33   (a) They are joined normally by the combined action of polymerase I and DNA ligase, neither of which are temperature-sensitive. (b) Not many. A few fragments being extended by polymerase I might have a few ^{3}H-thymidines very near the 3' terminus.

8-34   Yes, unless the RNA molecule is copied from both ends, copying only one strand from each end.

8-35   The gap will be filled by a single piece. No fragments would be formed as long as one end of the gap had a 3'-OH group.

8-36   Bidirectionality of replication.

8-37   (a) Its displacement and nick-translation activities.
(b) Its density would decrease because the single-stranded branch would be removed
(c) It would decrease.

8-38   I, negative; II, positive.

8-39   (a) One twist/200 base pairs = one twist per 20 turns of the helix. Similarly, 1 per 150 = 1-1/3 positive twists possible per 20 turns of the helix. Thus 2-1/3 turns of every 20 turns can be replicated or about 12 percent.
(b) Only about 4 percent of mitochondrial DNA is replicated when the D-loop reaches its maximum size. Thus, DNA replication probably cannot proceed once the DNA is overwound.

8-40   (a) Unwinding enzymes cause separation of a double helix. Helix-destabilizing proteins prevent separated strands from re-forming a double helix or a single-standed segment from self-annealing.
(b) In the replication of a circle the rotation of the replicating strands causes twisting of the nonreplicating portion. The ω protein cannot be involved in removing the twists because it can only remove twists in the opposite sense from those generated by replication. However, DNA gyrase can remove the twists.

8-41   Statements 1 and 2.

8-42   (a) The 3'-terminal base of the RNA primer is adenine.
(b) The first DNA base added to the primer is deoxyguanylic acid, that is, 5'...rApdG...3'.
(c) One can only conclude that deoxyguanylic acid is not the first deoxynucleotide added, showing that the primer does not have random ends.
(d) The ^{32}P ends up in the undigested DNA.

8-43   3'-OH.

8-44   (a) One.
(b) There is no maximum.
(c) One of the parental single strands.

8-45   Polymerase III cannot replicate from a nick so that something must occur to form a gap. However, this situation is no different from that at a growing fork in replication and the combined action of helicase and ssb protein should create the necessary conditions. From what we know about replication, polymerase I should be able to carry out rolling circle replication without polymerase III or helicase but that is probably not the case.

8-46   No, since it is unlikely to have a replication origin.

8-47   First, the parent strand is copied to make double-stranded DNA. Hence, the density shifts to the lower value for native DNA. Between 2 and 4 minutes the amount of ^{3}H-labeled DNA increases with respect to ^{14}C-labeled-DNA but without a change in density. Thus, this must represent replication of the linear helix to form another double helix. This process continues at least until 6 minutes, as the amount of ^{3}H-double-stranded DNA has increased. Note that at 6 minutes some ^{3}H-DNA has a slightly higher density. This is due to displacement replication—that is, only one strand of the double-helix is being copied (the one whose sequence is complementary to the ^{14}C-labeled strand), displacing a single strand. The displaced strands are released by 8 minutes. These are the progeny strands and presumably are packaged into phage particles. Note that the ^{14}C strand remains at the density of double-stranded DNA. Thus, the initial ^{14}C-containing helix is used only as a template for synthesis of more double helices and no ^{14}C ever appears in progeny phage.

8-48   (a) All molecules in the act of synthesis will be completed but no resting molecule will be started. Therefore, a population of cells randomly distributed throughout the growth cycle will continue to synthesize DNA for one generation.
(b) There will be no synthesis since, during the period of amino acid starvation, all molecules in the act of synthesis will be completed.

8-49   In media in which the cells grow rapidly, DNA molecules in the act of synthesis initiate a second round of replication before the first round is completed.

8-50   The rate of replication in *E. coli* is 2.6×10^9 molecular weight units or 4.2×10^{-3} picograms per 40 minutes or 1.05×10^{-4} pg/min. Since there are two forks in *E. coli*, the rate per fork is 5.2×10^{-5} pg/min. The rate in mammalian cells is 5 percent of that value or 2.6×10^{-6} pg/min. Thus, with one replication fork it would take $3/2.6 \times 10^{-6} = 1.15 \times 10^6$ min. Since replication is completed in 360 min, there must be 3200 forks. Since replication is bidirectional in mammalian cells, there are 1600 replication loops.

8-51

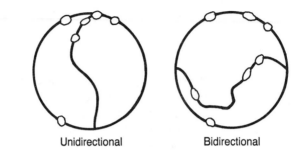

Unidirectional   Bidirectional

FIGURE A-8-51

8-52 No, since in both cases the replication forks have passed the point which is halfway around the circle.

8-53 All modes require sites for initiation of all replication and sites for initiation of synthesis of precursor fragments. Unidirectional replication also needs a stop signal to prevent extension of the first precursor fragment; bidirectional replication does not need any other sites; single–stranded circular DNA needs a region to which ssb protein is not bound and at which priming can occur.

8-54

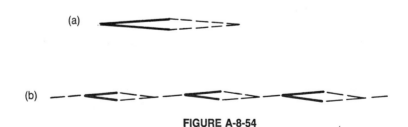

(a)

(b)

FIGURE A-8-54

8-55 The *F'* plasmid has probably integrated into the chromosome. Hence, strain C contains both alleles. The phenomenon could occur in strains carrying *dnaA*⁻ or *dnaE*⁻ mutations because these mutations are recessive.

8-56 The most recently synthesized DNA is probably membrane-bound.

8-57 Methylation of the bases to prevent self-destruction by restriction endonucleases.

Chapter 9

9-1 Replication errors, repair errors, loss of bases due to chemical instability, attack by environmental agents.

9-2 Spontaneous deamination of cytosine and depurination.

9-3 (a) Ultraviolet light.
(b,e,f) Ionizing radiation such as x rays and gamma rays.
(c) Alkylating agents and low pH.
(d) Psoralen, bifunctional alkylating agents, certain antibiotics, nitrous acid.

9-4 (a) Inhibition of cell division, damage to the RNA or protein-synthesizing system, damage to any essential system.
(b) Not likely if there are several such elements in the cell. For example, inhibition of the ability to synthesize proteins would require many "hits" because there are many copies of the protein-synthesizing system.

9-5 (a).

9-6 Provides activation energy for the dimer-cleavage reaction. Phosphorylation by ATP.

9-7 Statement 2 is true for both types of curves.

9-8 Use a set of filters that enable you to irradiate the cells with a particular range of wavelengths. Measure the effectiveness of killing (e.g., the dose yielding a particular survival value) as a function of wavelength. The curve showing this function is called an *action spectrum*. If DNA is the primary absorber, the action spectrum should match the absorption spectrum of DNA.

9-9 (a) Photoreactivation.
(b) A base-N-glycosylase.
(c) Excision repair.

9-10   PolI and ligase are essential for normal DNA replication. Therefore, the mutants, which are viable, must have a residual activity of these enzymes and this activity must be sufficient to carry out some excision repair.

9-11   Either the damage to T4 DNA is, for some reason, not reparable by the Uvr system or T4 possesses its own repair system. The latter is correct.

9-12   It is inducible and it is error-prone.

9-13   (a) The PR set and DR set of dimers are distinct.
       (b) 15 percent of the dimers are common to both sets.
       (c) same as (b).
       (d) 10 percent of the dimers are common to both sets.

9-14   B/r probably existed in the population before irradiation and was recovered since it had a higher probability of survival than the wild-type. This could not have been the case for B_S, which would not have survived. Therefore it seems reasonable to think that B_S arose from a mutation induced by the UV-irradiation. If this were so, the colony would be expected to contain wild-type cells, since parental cells are always produced in the first division after mutagenesis. However, this also seems unlikely, since the original screening test for B_S consisted of showing that the colony contained mostly sensitive cells; if 50 percent were resistant, the colony would not have been recognized as one containing mostly sensitive cells. Therefore, it seems likely that the parental DNA strand—that is, the strand containing the wild-type base sequence for the B_S gene—must have suffered additional damage (for example, a lethal mutation), so that it would not give rise to viable progeny. Other explanations can probably also be given.

9-15   If the dimer is in the strand being copied by the leading strand, only the leading strand would have a gap because the open helix stabilized by ssb protein allows synthesis of the lagging strand. The important question is whether a dimer is a block to helicase. If so, a dimer in the template for the lagging strand would block advance of the leading strand and produce gaps in both leading and lagging strands. There is no information at present on this point.

9-16   The secret in answering this question is to realize that trans-dimer synthesis may not require an RNA primer. Thus occasionally, a second round of replication might get started if a dimer were very near the terminus of replication and trans-dimer synthesis ensued past the replication origin.

9-17 If DNA replication occurs before sufficient repair of the damage is done, nonfunctional DNA might be formed. Also, excision of segments from both single strands could lead to double-strand breakage.

9-18 (a) X is not inducible, because the enzymes were present before protein synthesis was blocked by chloramphenicol.
(b) X would probably be considered inducible. The residual 5 percent could be due to either a second noninducible system, which can excise thymine dimers, or to a small amount of synthesis of X proteins when chloramphenicol is present.

9-19 In a *polA⁻* strain gap-filling cannot be carried out by the highly edited activity of polymerase I. Two other systems could fill the gap—polymerase III and the SOS repair system. Polymerase III is not particularly effective in filling small gaps nor can it carry out nick-translation, so presumably the error-prone SOS system does the job. In a UvrA⁻ cell thymine dimers are not excised so that SOS repair would also be more frequent and the mutation frequency would be higher for a given dose than with Uvr+ cells.

Chapter 10

10-1 Mutant—an organism whose phenotype differs from the norm; mutation—a base sequence differing from that in a wild-type organism; mutagen—an agent or reagent that causes mutations to occur; mutagenesis—the act of causing mutations.

10-2 Gal⁺ is a phenotype; *gal⁺* is a genotype.

10-3 Spontaneous mutagenesis.

10-4 A nonsense mutation stops chain growth at the mutational site; a missense mutation causes an amino acid substitution at the mutational site.

10-5 This mutant could not be isolated because there is no temperature at which it could grow.

10-6   Only bases 1 through 6 are shown.

(a) First round: One molecule each of

C A T T A A   C T T T A A
     and
G T A A T T   G T A A T T

Second round: Three molecules and one molecule, respectively of

C A T T A A   C T T T A A
     and
G T A A T T   G A A A T T

(b) First round: One molecule each of

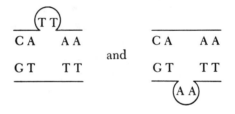

Second round: Two molecules and two molecules, respectively, of

C A A A   C A T T A A
     and
G T T T   G T A A T T

10-7   Several base sequences (codons) often correspond to the same amino acid.

10-8   The amino acid may not be in an active region of the protein. In addition, the exchange may not affect the conformation of the protein. Changes from polar to nonpolar amino acids and vice versa (from nonpolar to polar) or a change in the sign of the charge are very likely to lead to mutation. A change to or from a cysteine or a proline (the latter alters the shape of the polypeptide backbone) is invariably an effective mutation.

10-9   (1) Purple, (2) purple, (3) purple, (4) purple, (5) pink.

10-10   The original cell contains an altered base in one strand. Following the first round of DNA replication, which occurs while the bacterium is on the agar surface, two bacteria, one Lac$^+$ and one Lac-, result. Since they are adjacent on the agar, the colony has a Lac$^+$ half (purple) and a Lac$^-$ half (pink).

10-11   Add both sulfonilamide and penicillin and grow for a period of time. Penicillin will kill the Sul$^+$ bacteria but not the Sul$^-$ bacteria. Then, plate on agar lacking sulfonilamide and replica-plate onto agar containing sulfonilamide. Select colonies' that fail to plate on agar containing sulfonilamide.

10-12   The ratio of polymerizing activity to exonuclease function will decrease in the antimutator since errors will be removed more often.

10-13   A base change for which the phenotype is wild-type.

10-14   In the course of many rounds of replication in the United States and in France, silent mutations will accumulate. These mutations will not be the same, so that a U.S.-France heteroduplex will have mismatched pairs of bases and thus short single-stranded regions.

10-15   The strain has a leaky mutation in the *leu* gene; the strain grows but the mutation produces slow growth in the absence of exogenous leucine.

10-16   Many amino acid changes will not yield a mutant phenotype and many mutations will be chain-termination mutants.

10-17   The change must be in the DNA strand that is transcribed, and the change must be in a position within a codon that changes an amino acid.

10-18   (1), (3), (7), and (8), as they involve either changes in polarity, charge sign, or chemical properties.

10-19   A transition is a base-pair change in which the purine-pyrimidine orientation is not altered; a transversion is a base-pair change in which the orientation is changed.

10-20   NNG often produces clusters of mutations. Frequently, one of these mutations is either very leaky or in a nonessential but useful gene.

10-21   5-Bromouracil is a base analogue substituting for thymine and sometimes pairing with guanine; it induces transitions. 2-Aminopurine substitutes for adenine and guanine and can pair with both thymine and cytosine, inducing transitions. Nitrous acid causes deamination of guanine to xanthine, adenine to hypoxanthine, and cytosine to uracil. Hypoxanthine pairs with cytosine and uracil with adenine, so transitions are produced. Xanthine cannot pair with any of the standard bases. Acridine orange intercalates and produces frameshifts.

10-22   A replication error might arise if the 3' terminus of a growing strand became temporarily unpaired (owing to breathing) and a base was copied twice. This could happen if a single base was repeated sequentially and temporary misalignment of one strand with the other occurred after a breathing cycle. A recombination error could also arise by mispairing to produce an error in register in a repeated base sequence such as AAAAA or AGAGAGAG.

10-23   Addition of two bases or deletion of one base.

10-24   In the first case, measure the reversion frequency; in the second, perform genetic crosses with known mutants and deletions and determine the recombination frequencies with each.

10-25   Since the frequency is detectable, the mutant cannot be a deletion. However, the frequency is too low to be a point mutant. Thus, the mutant is probably a double mutant.

10-26   A frameshift mutant should revert following treatment with acridines. A nonsense mutation can be reverted if it is put into a cell containing an effective chain-termination suppressor. Since a particular suppressor may respond only to a particular chain-termination mutation and since the amino acid placed at the site of a chain-termination mutation may not always yield a functional protein, several suppressors must usually be tried. A missense mutant fails to respond to both of the preceding treatments.

10-27   No. The frequency of exact replacement should be at least 1000 times lower.

10-28   (a) 10^{-12}, the product of the reversion frequencies of each.
(b) A second frameshift mutation (*his3*) could counter the effect of both the *his1* and *his2* mutations. It would have to be of opposite sign from *his2* (that is, a deletion, if *his2* were an addition and an addition, if *his2* were a deletion). If such a second frameshift mutation were located on the side of *his1* opposite *his2*, it would restore the reading frame, and *his1* would lie in the short segment of altered reading frame where it would no longer be read as part of a termination codon. The second frameshift codon would have to lie close enough to the first one that essential amino acids in the protein would not be part of the altered segment.

10-29   They clearly interact. Since a charge sign-change yields a mutant and a second sign-change in another amino acid yields a revertant, amino acids 28 and 76 are probably held together by an ionic bond.

10-30   Mutagenize the parent strain and grow the mutagenized culture at 30°C in glucose-salts minimal medium containing GP. Wash the cells in a salt solution, and resuspend in a medium that is similar but contains ^{3}H-labeled GP. Allow the cells to grow at 42°C for several generations. The wild-type cells will incorporate ^{3}H-GP and the desired mutants will not. Wash the cells in a salt solution to remove unutilized GP and store at 5°C until ^{3}H decay has killed 99.999 percent of the cells, as determined by plating at 30°C. Test the surviving colonies for ability to grow at 42°C. Check the Ts mutants (that is, those which cannot grow at 42°C) for ability to incorporate ^{3}H-GP or ^{32}P-GP at 42°C into material which sediments with bacteria and qualifies chemically as phospholipid. The desired mutant should not incorporate GP into phospholipid.

10-31   Nitrosoguanidine produces mutations in clusters. Since *a* and *b* are nearby, with high probability a nitrosoguanidine revertant of an a^- mutation has other nearby mutations, many of which will be in gene *b*.

10-32   (a) MTG makes mutations in clusters. (See also problem 10-31.)
(b) Amino acid starvation prevents initiation of a round of DNA replication but does not inhibit replication that has begun.
(c) Re-addition of amino acids allows initiation to occur.
(d) Protein synthesis is needed to initiate a round of DNA replication.

10-33   The size, from the largest gene to the smallest, is *galC*, *galB*, *galA* because nonsense mutations are most likely to give a Gal⁻ phenotype for each mutational event (that is, each base change.)

10-34   The new *rII* mutation must be a deletion mutation which spans the gap between the *A* and *B* cistrons. It will eliminate parts of both genes but not the parts which contain the indicated point mutations.

Chapter 11

11-1   (a) Ribonucleoside 5'-triphosphates.
       (b) Double-stranded DNA.
       (c) RNA polymerase
       (d) No.

11-2   (a) PPP-5'-AGCUGCAAUG-3' and PPP-5'-CAUUGCAGCU-3'.
       (b) If the system required definite promoter sequences, probably no RNA would be made. If conditions were chosen for nonspecific initiation, both RNA molecules would be made.

11-3   The reactions are identical—namely, reaction of a nucleoside 5'-triphosphate with a 3'-OH terminus of a nucleotide to form a 5'-3'-dinucleotide. The substrates differ in that DNA polymerase joins deoxynucleotides and RNA polymerase joins ribonucleotides.

11-4   Events 2 and 3.

11-5   A 5'-triphosphate and a 3'-OH.

11-6   Core enzyme—the β β'α α tetramer; sigma factor—subunit responsible for recognition of the promoter; holoenzyme—core enzyme plus sigma factor; promoter—region of DNA containing RNA polymerase start sequences; operator—a repressor binding site.

11-7   This is the so-called "one gene-one enzyme" result; that is, a mutation in a particular gene never affects more than one enzymatic activity. In 1977 departures from this phenomenon were discovered in *E. coli* phage øX174.

11-8   At low ionic strength RNA polymerase binds at random to numerous charged sites rather than to specific promoter sequences. At high ionic strength the charges are shielded and binding occurs only to promoters.

11-9   The holoenzyme is needed to initiate RNA synthesis at a promoter. The core enzyme is needed for continued synthesis after initiation.

11-10   (a) DNA that is bound to protein is resistant to pancreatic DNase. Thus, if a complex of DNA and RNA polymerase is exposed to DNase, only the segment of DNA over which the bound RNA polymerase extends will survive enzymatic digestion. The number of nucleotides and the base sequence in this region can be determined by simple additional measurements.
(b) Bases in close contact with a protein are resistant to methylation. Thus, the bases in contact with the protein can be identified.

11-11   (a) An RNA molecule containing one or more contiguous base sequences that are translated into one or more proteins.
(b) A primary transcript is a complementary copy of a DNA strand. It may contain mRNA, tRNA, or rRNA and may be processed before translation can occur.
(c) A coding strand is a segment of a DNA strand that is copied by RNA polymerase. An antisense sense is a DNA strand that is complementary to a coding strand.
(d) Cistron is a DNA segment between and including translation start and stop signals that contains the base sequence corresponding to one polypeptide chain. A polycistronic mRNA molecule contains sequences encoding two or more polypeptide chains.
(e) Leaders, spacers, and the unnamed region following the last stop codon of a mRNA are untranslated regions.

11-12   A transcription unit is a section of DNA extending from a promoter to an RNA polymerase termination site. It is usually not a gene, but typically includes many genes.

11-13   Promoter-up mutation: a new promoter is created or the activity of an existing promoter is enhanced; promoter-down mutation: a natural promoter is inactivated.

11-14   Open-promoter complex—a complex between RNA polymerase and a promoter in which initiation can begin; closed-promoter complex—one in which initiation cannot begin. Presumably more base pairs are broken in an open-promoter complex.

11-15   The binding of RNA polymerase to DNA is sensitive to the number of superhelical twists. Possibly, binding is facilitated by the partial unwinding of a superhelix.

11-16   RNA polymerase must cause either a slight underwinding or overwinding of the segment of a DNA molecule to which it is bound. Whether it causes under- or overwinding would result in either a negative or a positive superhelix.

11-17   Actinomycin intercalates and generally inhibits transcription. Rifampicin inhibits initiation of RNA synthesis. Streptolydigin inhibits chain elongation.

11-18   The rates of initiation of synthesis of different mRNA molecules are not all the same.

11-19   The promoter sequence is first determined. The 5'-terminal sequence of the transcript is then obtained (4 to 5 bases are sufficient) and the complement to this sequence is found in the promoter sequence.

11-20   Upstream and downstream usually refer to regions in the 5' and 3' directions, respectively, from a particular site that is being discussed. Sometimes, the terms are used specifically in the following way: upstream—a region of the DNA before the first base of an mRNA that is transcribed; downstream—in the 3' direction (of the RNA) from the first base transcribed.

11-21   (a) It is probably a sequence that existed at very early times and from which others were derived by mutation. Furthermore, it indicates that the biochemical system that uses the sequence existed very long ago.
(b) It is essential for some stage of promotion—in particular, binding of RNA polymerase or initiating polymerization.
(c) The rates of initiation may differ slightly. It is difficult to be certain of this point though because the rate of initiation is mainly determined by the -35 region.

11-22   It is self-priming; it can unwind the helix locally without the aid of an auxiliary helicase.

11-23   Only a few base pairs are broken during transcription. The open region is smaller than the size of RNA polymerase, which is apparently able to break a few base pairs by itself, as it moves along the helix.

11-24   The Rho factor.

11-25   A sequence that can self-anneal to form a segment of double-stranded RNA, a sequence of AU pairs, and a (G+C)-rich region.

11-26   In transcription a time interval after RNA synthesis begins during which the polymerization rate is noticeably less than the average polymerization rate.

11-27   (a) The graph shows that 20,000 cpm does not decay. This is the stable RNA or 20/50 = 40 percent of the total.
(b) To obtain this number, subtract 20,000 from each value and replot on semi-log paper. The half-life is two minutes.
(c) These numbers can be used without correcting for stable RNA, so the half-life is six minutes.
(d) Uridine is ultimately converted not only to rUTP but to dUTP, which then forms TTP and is incorporated into DNA. The DNase treatment insures that all acid-insoluble radioactivity is in mRNA.

11-28   Since the triphosphate of the primary transcript is absent, the molecule has been processed. The processing could be extensive or as simple as triphosphate hydrolysis but there is no way of knowing from the information given.

11-29   Formation of (1) the 3'-OH terminus, (2) the 5'-P terminus, and (3) modified bases.

11-30   In eukaryotes the DNA is arranged in nucleosomes.

11-31   RNA polymerase I makes rRNA in the nucleolus. RNA polymerases II and III are nuclear; the former makes mRNA and the latter makes tRNA and 5S RNA.

11-32   (a) A terminal structure in which a methylated guanosine is in a 5'-5'-triphosphate linkage at the 5' terminus of mRNA.
(b) The 3'-OH end.

(c) All mRNA molecules (except those of several viruses) are capped. Some mRNA molecules lack the poly(A) tail.

11-33   (a) In eukaryotes all mRNA is monocistronic.
(b) Only the translation start signal nearest to the cap is utilized.

11-34   The collection of primary transcripts and partially processed transcripts in the nucleus.

11-35   (a) Untranslated sequences that interrupt the coding sequence of a transcript and that are removed before translation begins.
(b) Removal of introns.

11-36   Eight.

11-37   (a) *E. coli* is a prokaryote and its RNA polymerase does not recognize all eukaryotic promoters. Yeast, on the other hand, is a eukaryote.
(b) *E. coli* cannot remove the introns from the mRNA. Yeast can. This problem can be avoided if the transferred gene is prepared (by techniques described in Chapter 20) by copying a processed mRNA.

Chapter 12

12-1   Met Pro Leu Ile Ser Ala Ser.

12-2   (a) (c) (d).

12-3   Histidine, glutamine, cysteine, tryptophan, serine, glycine, leucine, proline, isoleucine, threonine, lysine, and methionine.

12-4   Arg to His;   Met to Ile; Gly to Asp.

12-5   (3 x 124) + 3 (termination) + 3 (initiation) = 378.

12-6   For UAG, the amino acids are Tyr, Leu, Trp, Ser, Lys, Glu, and Gln. For UAA they are Tyr, Lys, Glu, Gln, Leu, and Ser.

12-7   Fifteen. One needs one codon to separate coding sequences—it could be both a start and a stop.

12-8   Arg-2 could be replaced by Gly, Trp, Lys, Thr, or Ser. Arg-3 could be replaced by Ser, Lys, Thr, and Gly. The Arg-1 codon cannot be identified.

12-9   The number of base pairs is $2.7 \times 10^9/660 = 4.09 \times 10^6$ which corresponds to 1.36×10^6 codons. The number of amino acids per protein molecule is 454 so there would be about $1.36 \times 10^6/454 = 3000$ proteins.

12-10   (a) AUG, starting at the sixth base from the left.
        (b) It is not in the same reading frame as the AUG.

12-11   (a) It could be the C-terminal peptide.
        (b) The key to solving this problem is noting that the Met codon must be AUG and that shortening the peptide means that a chain termination codon has been generated. Since the terminal Glu codon in the wild-type sequence is GA_ and the Val codon is GU_, the simplest explanation for termination is that the Val codon is GUU and there is a single base insertion that joins the second U to the GA of the Glu codon. Since the Thr codons all start with AC and Met is gone, a C must be inserted to the right of the A in the Met codon.

12-12   Redundancy is important in this question—several changes may be possible. For example, UUU to UUA is a Phe-to-Leu change and UUA to UUC will restore Phe. Similarly, UGU to UGG yields Trp and a change to UGC restores Cys. Many mutagens can restore original base sequences.

12-13   ...Val-Cys-Val-Cys-Val-Cys... and peptides of various sizes starting with Val or Cys.

12-14   The frequencies of the triplets are: (3G), 1/8; (3U), 1/8; (2G,1U), 3/8; (2U,1G), 3/8. The (2G,1U) combination yields equal amounts of Gly (GGU), Val (GUG), and Trp (UGG). The (2U, 1G) combination yields equal amounts of Leu (UUG), Cys (UGU), and Val (GUU). GGG and UUU yield Gly and Phe respectively. The totals are (Val and Gly appears twice): Gly, 1/4; Trp, 1/8; Leu, 1/8; Cys, 1/8; Phe, 1/8; Val, 1/4.

12-15   In any sequence of three bases in which A occurs five times as often as C, the probabilities of having A or C in a particular position are 5/6 and 1/6 respectively. Therefore, the probabilities for each triplet are: AAA, $(5/6)^3 = 125/216$; ACA,

AAC, CAA, 5/6 x 5/6 x 1/6 = 25/216 for each ; ACC, CAC, CCA, 5/6 x 1/6 x 1/6 = 5/216 each; CCC, $(1/6)^3$ = 1/216. If the relative amount of AAA is defined to be 100, then the relative amounts of each of the other triplets are: ACA, AAC, CAA, 25/125 x 100 = 20 each; ACC, CAC, CCA, 5/125 x 100 = 4 each; 1/125 x 100 = 0.8. Therefore, the codon assignments are: Lys, AAA; Asn, (2A,1C); Gln, (2A,1C); His, (1A,2C); Pro, CCC and (1A,2C); Thr, (2A,1C) and (1A,2C). When the bases are given in parentheses, their order in the codon is not known.

12-16   (a) Arg, Asp, Thr.
        (b) Met, Asp.
        (c) Ile, Asn.

12-17   AAA, AGA, CAA, CGA, GAA, GGA, UUA, UCA UAC, UAU, UGC, UGG, UGU, UUG, UCG, CAG, GAG, AAG.

12-18   If it is very near the terminus of the protein so that only a few nonessential amino acids are removed.

12-19   (b).

12-20   Sometimes, if the reading frames are different. For example, one protein could be translated from the sequence...5' GGU-UCA-3' ... (the hyphens indicate the reading frame), whose complementary sequence is ...5' U-GAA-CG3' ... A change in the first sequence of C to A generates UAA in the reading frame. This change also generates UAA in the reading form of the complementary strands. Other arrays can also be changed to form two out-of-phase stop codons.

12-21   (a) If A were added, the terminating codons would be UUA, UAA, and AAA, in decreasing frequency. The amino acids attached to polyphenylalanine would be leucine and lysine (UAA is a stop codon). For C and G the amino acids are serine (UCC) and proline (CCC), tryptophan (UGG), leucine (UUG), and glycine (GGG). (b) No. Thirty-six triplets would not be determined.

12-22   Wobble refers to the possibility of forming base pairs other than AU and GC in the third position of a codon-anticodon pair.

12-23   Thirty-two.

12-24   (a) AUG might have to be near a ribosome binding site or some

other special sequence in order to serve as a start signal.
(b) A stop codon will not cause termination when it is not in the reading frame.
(c) It is the first AUG in the 3' direction from a Shine-Dalgarno sequence or the first AUG following, but very near to, a stop codon.
(d) The initiating AUG in eukaryotic mRNA is the one nearest the 5' terminus of the mRNA.

12-25   The Tyr and His codons followed by a termination (UAA) codon are easy to find in the mRNA. To add two more amino acids, one has to add a nucleotide to make two extra codons. This must be done in .such a way as to break up the termination codon, adding one base and generating codons for Leu and Ser. AGC codes for Ser, so something must be added to the U and A of the termination codon to get a Leu codon. The only way to do this is to add a U on either side of the U in the termination codon. Thus there must be a frameshift mutation adding U to give the sequence 5'-AAGUAUCACUUAAGC.

12-26   (a) The transcript from DNA strand 2 is 5'-X-AUA-UAG-GGG-GCA-Y-3'. The transcript from DNA strand 1 is 5'-Y'-UGU-CCC-CUA-UAU-X'-3'. The DNA strand 2 transcript contains an amber codon so that the coding strand is 1 and the amino acid sequence is Cys-Pro-Leu-Tyr-COOH.
(b) A single-step mutation of the UGC codon yields UGA, a nonsense codon; mutation of the UAU codon yields UAA or UAG, both nonsense codons.
(c) The mRNA sequence for such an amino acid sequence is

$$\text{UG} \begin{pmatrix} U \\ C \end{pmatrix} \text{CC} \begin{pmatrix} U \\ U \\ C \end{pmatrix} \text{UA} \begin{pmatrix} U \\ C \end{pmatrix} \text{AUG,}$$

where the symbols in parentheses represent alternative bases at one site, each of which would yield the same amino acid sequences. Given the base sequence of the wild type, the mutant sequence must be Y'-UGC-CCC-UAU-AUG, which cannot be generated from the original sequence by a single base substitution. Instead, a deletion of a C in position 3, 4, 5, 6, or 7 (starting with UGC) must have occurred.
(d) From (c) and since the third base of the Met codon is G, X' must be G and X must be C.

12-27   The G-to-C mutation creates a UAG stop codon. Thus the AUG found "downstream" (and in the proper reading frame) can start in-frame synthesis going again. Although not shown, there must be a ribosome binding site somewhere to the left of the sequence.

12-28   (a) A frameshift mutation has not only altered the protein but removed the stop codon of cistron *A* from the reading frame. Thus, the AUG of cistron *B* is not encountered.
(b) A stop codon that is in the new reading frame in *A* has been introduced. This allows the AUG of cistron *B* to start a new reading frame. The suppressor prevents chain-termination of the mutant *A*.

12-29   3 and 4.

12-30   1.

12-31   (a) Leucine.
(b) Serine.

12-32   Protein synthesis would occur, but all isoleucine positions would contain valine instead.

12-33

FIGURE A-12-33

12-34   Stem—a double-stranded region. Loop—a closed single-stranded region attached to a stem. Arm—a stem plus its attached loop. Tertiary base pairs—base pairs, not necessarily between the bases themselves, causing the three-dimensional folding of tRNA.

12-35   Base-base, base-ribose, and ribose-ribose bonds. The second and third type are used only for tertiary folding.

12-36   Binding is on only one side of the nucleic acid.

12-37   An aminoacyl-AMP must be available; the synthetase must be able to bind to the tRNA; the three-dimensional structure of the tRNA must be such that the CCA terminus is positioned in the active site of the enzyme.

12-38   A tRNA molecule that either can "read" a chain-termination codon or, with low frequency, either misread a codon or be mischarged. Such molecules arise from mutations in genes for tRNA molecules.

12-39   There are two possibilities: (1) There are two genes for the original tRNA species, and only one of these has been mutated. (2) Natural chain-termination sequences might usually consist of two or more different termination codons. The nonsense suppressor will suppress only one, and chain termination will still occur. Both possibilities occur.

12-40   UGA is usually preceded or followed by another chain termination codon.

12-41   Binding to a UUA anticodon must be weak. This is necessary because UAA often occurs singly as a natural stop codon.

12-42   Some possibilities are (1) recognition of the wrong tRNA by a mutant aminoacyl synthetase, (2) mischarging by a tRNA mutated in the recognition loop, (3) recognition of a wrong codon by a tRNA mutated in the anticodon itself or in the anticodon loop next to the wobble base in such a way that the codon-anticodon specificity is reduced. Missense suppressors must be weak in order that all normal proteins do not become altered. A Y-for-X missense suppressor need not be active against all X-type mutants because substitution of Y may not always result in a functional protein.

12-43   The products of genes a and b mutually interact. The mutational changes in each are in the interaction sites and are compensatory.

12-44   A tRNA molecule having a four-base anticodon could suppress a base addition. Such a mutant tRNA molecule could be produced by acridine mutagenesis since the acridines produce base additions. A single base deletion would have to be suppressed by a tRNA molecule having a two–base anticodon. This is not impossible but seems unlikely because, if it occurs, many of the normal tRNA molecules having three-base anticodons would be able to suppress the deletion and, therefore, the deletion would not be picked up as a mutant. On the other hand, perhaps such suppression does occur, but, because suppression could be very weak and because so many different amino acid substitutions would occur at the mutated site, the number of active protein molecules would be so low that reversion might not be detectable.

12-45   To select a temperature-sensitive amber suppressor, select a bacterial strain that has an amber mutation in gene Y and an amber suppressor. Grow the strain at high temperature in a growth medium lacking Y and containing penicillin. When this antibiotic is present, prototrophs will die and auxotrophs will survive. A mutant with an amber suppressor inactive at high temperature will fail to suppress the amber mutation and, thus, the mutant will survive the penicillin treatment.

12-46   (a) The amino acid replacement is not the original amino acid and the new protein functions only at the lower temperature. (b) Amino acid C normally interacts with A. Neither C and B nor A and D interact, but B and D can interact. Thus, C elsewhere in the protein is replaced by a D so that the protein can fold properly.

12-47   Both types are suppressed because of wobble—the anticodon for UAA is UUA and the first anticodon U can, by wobble, pair with a G in the third position of the codon; thus, the anticodon UUA can pair with the codon UAG. Since a UAA codon can also, by wobble, be suppressed by a tRNA with the IUA anticodon, which cannot pair with a UAG codon, such a mutant could account for the "restricted"-range suppressor. The amino acids is one whose tRNA molecule could be mutated to have an IUA anticodon. The amino acid must have a codon starting with UU, UC, UG, CA, GA, or UA, and be capable of having an I in the anticodon (that is, be at least triply redundant). Only serine (UCA, UCG, UCU, UCC) has this property.

12-48   The rIIA and rIIB proteins are not translated in the same absolute reading frame—there are fourteen nucleotides between the C-terminal Ala of the rIIA protein and the N-terminal fMet of the rIIB protein. Any deletion fusing rIIA and rIIB peptides must put them into the same frame. The deletion must take out $14 + 3n$

nucleotides, in which n is any integer. The values $984 - 14 = 970$ and $988 - 14 = 974$ are not multiples of 3 and must be wrong; the values $980 - 14 = 970$ and $1088 - 14 = 1074$ are multiples of 3 and could be right.

12-49 (a) If the stretch of DNA corresponding to genes 9 and 7 is inverted and reinserted as described, the antisense strand of this segment must become connected to the sense strands of genes 8 and 5 to preserve polarity, as shown in Figure A12-49. Since no new promoter is present in the mutant, the entire cluster will continue to be transcribed from left to right into a single mRNA, in which the segment corresponding to genes 7 and 9 will carry not the normal message, but its complement in reverse order instead. Therefore, the enzymes coded by genes 7 and 9 will not be produced, and the mutant will be His⁻.
(b) The reversed antisense strand of the $9 + 7$ region which is not normally transcribed, may by chance contain a sequence that signals chain termination for mRNA synthesis. If so, gene 4 would not be transcribed.
(c) Regardless of how it is spliced into the mRNA molecule, the reversed complementary message transcribed from the $9+7$ segment is almost certain to include, by chance, one or more of the termination codons UAG, UAA, or UGA. The resulting polypeptide-chain termination will reduce the production of all the distal proteins, that is, 5, 4, 6, 3, and 2.

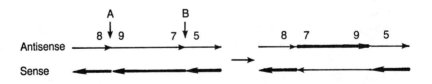

FIGURE A-12-49

Chapter 13

13-1 Prokaryotic: 70S with 30S and 50S subunits containing 5S, 16S, and 23S RNA. Eukaryotic: 80S with 40S and 60S subunits containing 5S, 5.8S, 18S, and 28S RNA.

13-2 Spontaneously.

13-3   The *L* proteins are components of the large (50S or 60S) subunit; the *S* proteins are components of the small (30S or 40S) subunit.

13-4   (a) They have extensive double-stranded regions.
(b) The RNA in the ribosome is also extensively double-stranded.

13-5   No. The protein may be required for the overall three-dimensional shape of the ribosome.

13-6   Phage RNA has a ribosomal binding site and can stick to a ribosome; rRNA cannot.

13-7   Steps 1, 2, and 4.

13-8   (a), (b), (c).

13-9   (2), and possibly (3) and (4).

13-10   (a) The amino terminus is made first.
(b) The 5' end is read first.
(c) Start—AUG; stop—UAA, UGA, UAG.

13-11 NH_2-Arg-Leu-COOH, when reading starts exactly at the 5' terminus. Reading can begin at other positions also yielding NH_2-Gly-Tyr-Arg-Lys-COOH and NH_2-Val-Ile-Gly-Lys-COOH and fragments thereof.

13-12   To provide subunits to initiate translation of another mRNA molecule and to initiate reading in the correct reading frame.

13-13   RNA polymerase to transcribe mRNA and tRNA, a few enzymes for processing tRNA, about 20 tRNA synthetases, several enzymes for regenerating GTP, one or a few for forming a peptide bond, one or a few for releasing a finished polypeptide and a few for protein processing—perhaps 40 to 50 enzymes in all. This represents about 2 percent of the *E. coli* enzymes.

13-14   Synthesis would have to await the completion of a molecule of mRNA. In the existing system protein synthesis can occur while the mRNA is being copied from the DNA. Thus, protein synthesis can start earlier than would be possible with reverse polarity, and the mRNA is relatively resistant to nuclease attack.

13-15 (a)(3) Label is found at the carboxyl end only. This experiment was originally used to determine the polarity of protein synthesis.
(b) (2) Label in amino acids at all positions.

13-16 Formylmethionyl-tRNA and mRNA plus (perhaps) other things are required *in vivo*; a high Mg^{2+} concentration is required *in vitro*.

13-17 An initiator tRNA responding to the GUG codon has been made in the infected cell. Possibly the tRNA responding to AUG has been inactivated also. There is no termination codon in this reading frame so the peptide is not released.

13-18 Statement 2.

13-19 (a) Met Lys Lys Lys ....
(b) Met Lys, because both P and A sites can be filled without the occurrence of translocation.

13-20 Formation of the 70S initiation complex and translocation.

13-21 Initiation factors are used to prevent interaction of the ribosomal subunits and for binding of tRNA to its ribosomal binding site; elongation factors are used to move tRNA from one ribosomal binding site to the next; termination factors are needed for release of the finished polypeptide chains and of the mRNA from the ribosome.

13-22 399 molecules—200 to activate 100 amino acids, 198 to form 99 peptide bonds and 1 for initiation.

13-23 Release of finished proteins from the ribosome.

13-24 Degradation of mRNA.

13-25 It is not a single protein but is a catalytically active site formed from portions of many proteins.

13-26 In eukaryotes the initiating amino acid is Met rather than fMet and there is no Shine–Dalgarno sequence; instead the AUG nearest the 5' terminus of the mRNA is the initiating codon.

13-27   Penicillin interferes with bacterial cell wall synthesis and therefore has no effect on eukaryotes. Actinomycin affects transcription in both bacteria and mammals and hence is not of clinical usefulness. The remaining antibiotics affect translation, but tetracycline and erythromycin cannot inhibit translation in mammalian cells. Thus, these two drugs are useful against infection. Puromycin and chloromycetin are active against both bacterial and mammalian ribosomes. However, in cases in which an infection is becoming dangerous to a patient, chloromycetin can be used but the toxic effects have to be dealt with in other ways.

13-28   Yes, and a polysome results when that occurs.

13-29   Less, because with polyribosomes several protein molecules can be synthesized simultaneously from a single mRNA molecule.

13-30   They are nearly the same—17 amino acids polymerized per second and 55 nucleotides or 18 codons synthesized per second.

13-31   Such a protein has 5000/110 = 455 amino acids so it is synthesized in 455/17 = 27 seconds.

13-32   The region from 45 to 60.

Chapter 14

14-1   Catabolic—degradative; anabolic—biosynthetic; metabolic—catabolic + anabolic + all else.

14-2   Positive—the active regulatory protein enhances transcription above what it would be in its absence (for example, cAMP bound to CAP protein). Negative—the active regulatory protein blocks transcription.

14-3   A cell in which a segment of the DNA exists in duplicate. The segment is usually a small fraction of the total DNA. The cell is diploid for all elements in the duplicated region.

14-4   Repressed—turned off, inactive; induced—turned on, active; constitutive, not regulated; coordinate—regulated together.

14-5 Regulation is transcriptional and exerted at the initiation and termination (attenuation) steps.

14-6 The true inducer binds directly to the repressor, whereas a general inducer may cause induction in any way. A gratuitous inducer stimulates transcription but is not a substrate of the enzymes of the operon; thus, it does not serve a useful purpose to the cell.

14-7 There is a mutation near the terminal region of cistron A giving rise to a new and highly active promoter from which mRNA synthesis can occur under conditions of repression.

14-8 (a) When X is lacking from the medium, Xase is without value. Cells in which 20 percent of their protein is useless must grow slightly more slowly. If the culture medium does not limit growth for other reasons, growth will be limited by how fast the necessary quantities of the required proteins can be made. The repressorless constitutive strain should grow 20 percent slower; that 20 percent disadvantage will be compounded in each generation. When the inducible strain has undergone 30 doublings, the constitutive strain will have undergone only $0.8 \times 30 = 24$ doublings. At that time, for every 2^{30} cells of the inducible strain, there are only 2^{24} cells of the constitutive strain, that is, 2^6 times as many cells of the former as of the latter. Therefore the ratio of wild-type (inducible) cells to repressor-less (constitutive) cells is $2^6 = 64$.

(b) If there is a preferred carbon source also present, the answer is essentially the same as in (a). If there is no other carbon source but X, the constitutive strain is no longer at a disadvantage.

(c) An uninducible mutant, owing to a repressor that is defective in binding the gratuitous inducer. The reasoning is the following. The gratuitous inducer is not a substrate of Xase, i.e., not a carbon source. Ability to remain uninduced would confer a growth advantage on a cell in these special conditions, so, if such a mutation arose spontaneously, it would be selected in these special circumstances.

14-9 Regulation by repression would be very unlikely since these systems are always turned on. It seems likely that the enzyme concentration would be regulated (perhaps autogenously) so that just the right amounts would be made.

14-10 Messenger RNA is made shortly after addition of lactose and this is followed by enzyme synthesis. This continues for two generations, at which time depletion of lactose causes establishment of repression, turning off of mRNA synthesis and hence of enzyme synthesis. Enzyme activity persists since β-galactosidase is fairly stable, although the activity per cell decreases as the cells divide.

14-11 Synthesis of *lac* mRNA stops shortly after glucose is added. Enzyme is made for several minutes until *lac* mRNA is degraded. Enzyme activity is unchanged because the enzyme is stable. After one generation the glucose is consumed and synthesis of mRNA begins.

14-12 2 and 3.

14-13 (c).

14-14 Two. One if glucose is present.

14-15 When the *lac* operon enters the female, no *lac* repressor is present so that transcription of the *z* and *y* genes occurs and β-galactosidase is made. However, shortly after transfer, repressor mRNA, and hence, repressor, are made. This will shut off transcription of the *z* and *y* genes, and hence, will shut off enzyme synthesis. If the female has the genotype $i^+o^cz^-y^-$, repressor is present and no enzyme is made. The o^c mutation in the female is irrelevant.

14-16 At very low levels all inducible operons are transcribed—for example, a repressor might come off the operator for an instant and RNA polymerase will get on. Hence, each cell will contain a few β-galactosidase and permease molecules. The permease will bring in a few lactose molecules and these will be converted to the true inducer, allo-lactose; then derepression can occur.

14-17 (a) Yes, I, yes.
(b) Yes, I, yes.
(c) Yes, C, yes.
(d) Yes, I, yes.
(e) Yes, C, yes.
(f) Yes, I, yes.
(g) No, neither, no.
(h) Yes, I, yes.

(i) No, neither, no.

(j) Yes, neither, no.

14-18   (a) Constitutive for both.

(b) Constitutive for β–galactosidase and inducible for permease.

(c) Constitutive for both.

(d) Constitutive for β–galactosidase and inducible for permease.

14-19   The repressor cannot bind inducer. No, because the constitutive component makes no enzyme and the inducible part cannot be induced.

14-20   (a)   **TABLE A-14-20**

	Expected results	
Experiment	Induced	Uninduced
5	100	0.1
6	100	0.1
7	100	100
8	100	100
9	0.1	0.1
10	0.1	0.1

(b) There is twice the amount of β–galactosidase activity because there are two z genes, both making enzymes.

(c) The most probable reason is that the $z1^-$ mutant protein is reducing the activity of the good z protein by negative complementation. In positive complementation, two defective proteins complement each other to give a good, or semigood protein. In negative complementation, a defective protein complements a good protein in a negative manner; that is, a defective one makes a good protein bad. A careful examination of experiments 5 and 8 should convince you that only when the $z1^-$ allele is induced is the activity of the z^+ protein reduced. Negative complementation can occur because β-galactosidase is a tetramer consisting of four identical polypeptide chains. In a diploid there is free mixing of the subunits made from each lac region. If one defective z^- peptide chain comes together with three other good z^+ peptide chains, it "poisons" the good chains and reduces their activity.

(d) The explanation is positive intragenic complementation; when defective subunits $z2^-$ and $z3^-$ mix to form a tetramer, they make each other "good".

14-21 (a) The small plasmid which carries the *lac* operator sequence is present in 20 to 30 copies per cell. There are few molecules of *lac* repressor in each cell. The 20 to 30 "extra" operator sequences which are not situated in a *lac* operon compete with the chromosomal *lac* operator sequence for the binding of *lac* repressor, resulting in at least partial induction.
(b) A "promoter-up" mutation in the *lacI* gene which generated overproduction of *lac* repressor could produce such a result.

14-22 There might be a mutation in the gene for adenyl cyclase or for the cyclic AMP receptor protein. Another possibility is a membrane transport mutant, if these sugars were to utilize the same transport system.

14-23 Since the repressor is a multisubunit protein, a mutant repressor that could not bind to the operator and could bind tightly to a good repressor subunit, thereby inactivating it, would be constitutive and *trans*-dominant.

14-24 The two proteins are made from the same polycistronic message. The rate of production of each polypeptide chain is proportional to the product of its mRNA lifetime and the rate of initiation of protein synthesis. Thus, the ratio of the rates of production of a β-galactosidase chain and a transacetylase chain is (90 sec x $1/3$ sec^{-1}) divided by (55 sec x $1/16$ sec^{-1}) or 8.8.

14-25 RNA is hydrolyzed in alkali to short oligonucleotides, which do not sediment. If the fractions had been treated with acid and filtered to collect acid-insoluble radioactivity, no material would have been found at the meniscus and the biochemist would have realized that something was wrong.

14-26 It must be a protein if it is inactivated by a chain termination mutation. Suppression by both an amber and an ochre suppressor indicates that the mutant is a chain termination mutant.

14-27 It is a positive regulator activated by an inducer. The *b1* mutant cannot be activated. The *b2* mutant is activated without an inducer.

14-28 Site a is an operator, b is the structural gene, and c is a regulator; a^- is an operative-constitutive mutant.

14-29 A deletion removing the region $z—tsx$ has resulted in fusion of the y gene to the pur operon. Transcription of the y gene no longer proceeds from the lac promoter, but from a promoter in the pur operon that is regulated by the pur operator.

14-30 (a) The PhoR protein, because in its absence the expression is insensitive to the presence or absence of phosphate. (b) The PhoB protein; it is a positive effector. The PhoB protein must be binding to DNA and acting as a positive effector necessary for synthesis of APase. If the PhoR protein were acting directly on the DNA, then it would have to be a repressor. Mutants of the $phoR$ gene do show constitutivity, as repressor mutants do, but a $phoB^-phoR^-$ double mutant would also show constitutivity if the $phoR$ gene were a direct repressor, and they do not. Therefore, since the PhoR protein cannot be a direct repressor of transcription, the PhoB protein must be a direct positive control element affecting transcription of the $phoA$ gene.
(c) If the PhoB protein binds to the DNA at a promoter-like spot, and the PhoR protein in the presence of phosphate acts to reduce the synthesis of the PhoA protein, then the PhoR protein probably forms a complex with the PhoB protein and phosphate, making it harder for the PhoB protein to bind at the promoter site. Thus, transcription is inhibited. The $phoB^-$ mutation allows only five percent expression of the $phoA$ gene in the absence of phosphate. In the presence of a $phoR^+$ allele the residual expression is still sensitive to phosphate expression (the expression is reduced to one percent) but in the presence of the $phoR^-$ allele the five percent expression of APase cannot be reduced by phosphate.

14-31 (a) Three kinds of repressor mutations might occur: (1) a repressor that cannot bind X—the operon is on; (2) a repressor that cannot bind to operator even when X is present—the operon is on; (3) a repressor that binds to the operator without X—the operon is off.
(b) Phenotypes of partial diploid with wild-type and each of the mutants are: first mutant, inducible; second mutant, inducible; third mutant, unable to synthesize X under all conditions.

14-32   (a) Two—CAP-cAMP and RNA polymerase.
        (b) None.
        (c) Two (repressor and cAMP-CAP).

14-33   Lactose is cleaved to glucose and galactose so the *gal* operon is
        derepressed. The presence of glucose depresses the cAMP
        concentration so that transcription is from *S2*. If the amount of
        lactose were limiting, when the glucose is depleted,
        transcription would be from *S1* and would occur if there were
        sufficient galactose to maintain the derepressed state. This
        would probably be the case since the concentration of the *gal*
        enzymes is not very high when the *S2* promoter is used. Thus most
        of the galactose is metabolized after the *lac* operon is
        derepressed, when lactose is limiting. If there is ample lactose,
        the *gal* operon will never be operating at maximal level.

14-34   They might be, since certainly there is no need to degrade the
        amino acids when glucose is present.

14-35   All are Ara+.

14-36   The cAMP concentration increases and CAP-cAMP forms. CAP-cAMP
        binds to the *ara* operon and transcription of the *araC* gene
        begins. The *araC* protein (P_r form) is made and arabinose binds to
        it converting it to the active P_i form, P_i binds to *pBAD* and
        transcription of the *araBAD* genes begins.

14-37   A transcription termination site whose terminating activity is
        regulated and which is not far downstream from the promoter.

14-38   (a) Anthranilate synthetase is the product of gene *E*; tryptophan
        synthetase is the product of genes *B* and *A*. The order of
        transcription is *E* to *A* as shown. Translation is coupled to
        transcription, that is, a polypeptide chain starts to be
        synthesized as soon as its conjugate stretch of mRNA appears.
        Thus, the polypeptide chain coded by gene *E* will appear before
        the polypeptide chains coded by genes *A* and *B*.
        (b) The operon is 6700 nucleotide pairs long. Thus, the time is
        6700/25 = 270 sec = 4.5 min.
        (c) When IPA gets washed out, repression sets in again. The last
        RNA polymerase molecules to transcribe the *trp* operon reach genes
        B and *A* about 3 to 4 minutes after they leave the promoter. This

means that the last bit of gene *B* and *A* mRNA is made later than the last bit of gene *E* mRNA.

(d) IPA must bind to the aporepressor without causing the conformational change that converts the aporepressor to an active repressor.

14-39   The glucose is irrelevant.
(a) One.
(b) None.
(c) One.

14-40   Post-translational modification is needed. Some possibilities are the following: (1) The active enzyme might be a multisubunit protein and the monomer concentration takes time to reach a value at which the equilibrium shifts toward aggregation. (2) The protein must be modified by cleavage or addition of a sugar (i.e., it is a glycoprotein) before it is active. (3) A metal ion has to be bound.

14-41   (a) a, unchanged; b, unchanged; f, decreased; c, decreased; d, decreased. The polar mutation affects only genes downstream from the promoter.

(b) It is a *trans*-dominant superrepressor mutant which does not undergo any allosteric change in the absence of biotin and therefore does not come off the DNA at the operator site when biotin is present.

(c) The bacterium seems likely to contain a naturally occurring mutation of the type described in part (b).

14-42   In the stringent response, rRNA synthesis is inhibited by amino acid depletion; in the relaxed response, rRNA synthesis continues unabated.

14-43   (a) Since there is no protein synthesis the concentration of charged tRNA increases and rRNA synthesis continues. Thus, inhibition of protein synthesis is not a prerequisite for the stringent response.
(b) Same as (a).

14-44   (a) G inhibits 5; H inhibits 7; G and H together or E alone probably inhibit 3: J inhibits 8; enzyme 1 could be inhibited by (G, H, and J), (E and J), (G, H, and C), or (C and E); the

( ) enclose the products that act together.

(b) Step 1, three isozymes; step 3, two isozymes.

Chapter 15

15-1   The bacteria have stopped growing so phage growth is not possible.

(a) Small plaques will usually result if the burst size is especially small or if the phage adsorbs very slowly to bacteria. If the phage particle is very small, the plaque will usually be large because smaller particles diffuse further in agar.

(b) If a phage is adsorbed to a bacterium before dilution into agar, new phage will be produced sooner than if there is a time delay before a phage comes in contact with a bacterium in the agar. Since there is only a limited time in agar in which phage can be synthesized (until bacterial growth stops), the size of the plaque depends upon how quickly a phage particle adsorbs to a bacterium.

15-2   (c) The halo is caused by lysis of some of the bacteria surrounding the plaque, the result of lysozyme which is released by cells within the plaque and which diffuses outward.

15-3   T7 has no tail and diffuses more rapidly from the site of initial infection than does T4. Thus, soon after the second cycle of infection, T7 phage particles are quite far from the cell that was initially infected.

15-4   Four to five.

15-5 3.5×10^{10}, which is calculated from tube D. The number in tube E is small and statistically less reliable. The numbers fail to follow dilution, because the number in tube E is statistically unreliable—the number in tube C is so large that many plaques must overlap and two plaques will be counted as one.

15-6   T4 will be turbid. T4*h* will be clear.

15-7   The cell culture lyses visibly at 60 minutes. This time is 20 minutes after the moi becomes greater than 1 phage per cell. An approximate outline of the situation can be obtained by assuming that the bacteria divide in steps (see Table A-15-7). The 100 colony-forming mutants are T7-phage-resistant.

TABLE A-15-7

Time, min	Bacteria per ml	Phage per ml
$t = 0$	2×10^7	5000
$t = 20$		$200 \times 5000 = 10^6$
$t = 30$	$\sim 4 \times 10^7$	
$t = 40$	All bacteria are infected	$200 \times 10^6 = 2 \times 10^8$
$t = 60$	Lysis	

15-8   After adsorption, divide the mixture in half. Add antiserum to one half to kill unadsorbed phage. Measure the number of infected cells by their ability to produce phage; this is the number of infective centers. To the other half, add chloroform to kill infected cells and measure unadsorbed phage. The multiplicity of infection equals

(Phage added - Unadsorbed phage)/Infected cells.

15-9   From Poisson's law (Appendix D), e^{-2} or 0.135 of the bacteria are infected. Therefore, the number of plaques is 200(1 - 0.135) = 173.

15-10   Use Poisson's law (Appendix D). The fraction of cells infected by two or more phage $= 1 - P(0) - P(1) = 0.8$. Therefore 80 percent of 10^8 bacteria ($= 8 \times 10^7$ bacteria) are infected by two or more phage.

15-11   The Poisson zero value is to be 0.1. Thus, $e^{-n} = 0.1$, in which n is the ratio of phage to bacteria. Thus, $n = 2.3$.

15-12   The lysate contains four types of phage—normal T2, normal T2h, T2 DNA in a T2h particle (that is, one having a T2h tail), and T2h DNA in a T2 particle. Only those with a T2h structure can adsorb to B/2 but a plaque can form only if the DNA is also T2h. All phage can plate on B.

15-13   The original lysate contains nonviable particles which can complement the defect in R2p⁻ but not in R2q⁻.

15-14   (a) These are bacteria to which no phage adsorb. Since the total moi = 5, the fraction of cells which is uninfected is e^{-5} = 0.007.

(b) This number includes bacteria to which P4, but not P2, adsorb, as well as the infected cells. The moi of P2 is 3; the fraction of cells getting no P2 is e^{-3} = 0.05. Since 0.007 get neither P2 nor P4, then 0.05 - 0.007 = 0.043 get one or more P4 but no P2. Hence $0.043 \times 10^8 = 4.3 \times 10^6$ bacteria get P4 but no P2.

(c) These are the bacteria to which P2 adsorb but no P4 adsorb. Using the above reasoning, e^{-2} = 0.135 get no P4, and 0.007 get neither; 0.135 - 0.007 = 0.128, and 1.28×10^7 cells adsorb P2 but no P4.

(d) These are the bacteria infected by at least one P2 and one P4. The number of cells getting at least one P4 and one P2 is $10^8 - (1.28 \times 10^7) - (4.3 \times 10^5) = 8.22 \times 10^7$.

15-15   Adsorption, entry of nucleic acid, transcription production of phage proteins, production of phage nucleic acid, assembly of particles (including encapsidation of nucleic acid), release of progeny particles.

15-16   A protein coat to enclose the nucleic acid , a component capable of adsorbing to a bacterium, an element designed to enable the nucleic acid to penetrate the cell wall. The latter could be an injection system or a component that stimulates the bacterium to take in the particle.

15-17   By destruction of host DNA; by conversion of host RNA polymerase to a form which cannot read host promoters; by inactivation of host RNA polymerase and replacement with phage-coded polymerase.

15-18   Lysis would probably occur before many (or any) phage were made.

15-19   (T3, T7); T1; T5; (T2, T4, T6); the parentheses group the T phage species having DNA with the same molecular weights.

15-20   Single-and double-stranded DNA (linear or circular); single-stranded RNA and segmented double-stranded RNA (linear only).

15-21   A phage gene is considered nonessential if its product is not needed for growing the phage in the laboratory. Those genes that duplicate host genes may be truly nonessential, even in nature.

15-22   Phage–host specificity refers to the fact that a particular phage can grow on only one or a small number of bacterial species or strains. The most frequent cause is the inability of a phage to adsorb, except to certain bacteria.

15-23   There are several approaches, each of which involve isolating phage mutants that eliminate the activity of X. It is also necessary to have an *in vitro* assay for X; this might be enzymatic or immunological, or might require isolation of the protein. First, consider a heat–sensitive mutant. With this, it can be shown that at high temperature, enzymatic activity or perhaps immunological reactivity (not as good a test) is substantially reduced. Second, consider a chain termination mutation. Here it can be shown by sedimentation or by gel electrophoresis that the protein has a lower molecular weight than the wild-type protein. Note the greater power of the second approach. With the heat–sensitive mutant, one merely correlates a property *in vitro* with that *in vivo*, but with a chain-termination mutant one observes not only the absence of the wild-type protein but the appearance of a new and smaller molecule.

15-24   The best way is to use a temperature-sensitive mutant and raise the temperature of the infected cell at various times. If the gene product is needed only in the interval between t_1 and t_2, then raising the temperature later than t_2 will not affect phage development and a period of heating at any time between $t = 0$, and $t = t_1$, followed by cooling, will usually not affect phage development.

15-25   The Sup^+ host supports the growth of a phage having a conditional mutation.

15-26   T4 DNA is cyclically permuted, it contains 5-hydroxy-methylcytosine instead of cytosine, it is glucosylated, and it is 4.5 times larger than T7 DNA.

15-27   (a), (d), and (e).

15-28   In a simultaneous infection the T4 nuclease destroys the T7 DNA.

If T4 infection occurs ten minutes after T7 infection, the T4 DNA is not transcribed because T7 has inhibited the *E. coli* RNA polymerase. Therefore, T7, but not T4 phage, are produced.

15-29   To prevent incorporation of cytosine into T4 DNA.

15-30   The T4-induced nucleases will destroy all newly synthesized phage DNA molecules since, in cases (a) and (b), glucosylation cannot occur.

15-31   No, because the DNA is terminally redundant and the injection sequence differs for each particle.

15-32   The enzymes work catalytically, the structural proteins (of head, tail, tail fibers, and so on) are stoichiometric. Regulatory gene products are usually stoichiometric.

15-33   Most likely, the genes were once acquired from a host cell by some recombinational event. The most likely explanations for their continued existence are either that they increase the burst size (at least when the phage are growing in nature), or that their hosts in nature make these enzymes either in low concentration or not at all.

15-34   Modify *E.coli* RNA polymerase to prevent recognition of promoters that have been used and are no longer used and to permit recognition of the next promoter that is needed.

15-35   (1), (3), and (4).

15-36   Yes. If the headful rule is followed, the deletion will create terminal redundancy.

15-37   The early transcripts always contain genes for DNA replication and genes that (in varied ways) regulate transcription. Late mRNA generally encodes the structural protein genes and the lysis system. T7 is an exception in that the replication and structural protein genes are initially on a single transcript.

15-38   T4 successively modifies RNA polymerase to recognize new promoters; T7 makes its own RNA polymerase, inactivates the *E. coli* enzyme, and uses slow injection to delay late transcription; λ alters the ability to ignore transcription termination signals.

15-39   The molecular weights of *E. coli* and T7 DNA are 2.7 x 10^9 and 2.5 x 10^7 respectively. Thus, a T7 DNA molecule is replicated in $(0.0025/2.7)$ x $40 = 0.35$ minutes. T7 replication begins 8 minutes after infection and lysis occur at 25 minutes. Hence, there is sufficient time for $25/0.35 = 51$ doublings. If T7 DNA replicated continuously and exponentially, there would be $2^{50} = 10^{15}$ DNA molecules, rather than a few hundred, at the time of lysis. Thus, if the number of molecules successively doubles, only 8 doublings can occur, which would take 2.8 minutes in total; thus, on the average, there must be about 1.8 minutes between rounds of replication. Alternatively, replication does not proceed by doubling.

15-40   The structural proteins from the late transcripts are needed in stoichiometric amounts. The early proteins are mostly enzymes needed in tiny amounts. Since both early and late mRNA molecules have nearly the same half-lives, more copies of late mRNA are needed than early mRNA.

15-41   Injection of T5 DNA proceeds in two steps. In the first step 10 percent of the DNA is infected using energy from the phage particle; injection of the remaining 90 percent of the DNA requires cellular energy. Furthermore, T5 DNA is unique in sequence (nonpermuted) and a particular end is always injected first.

15-42   Joining of cohesive ends is used by the so-called lambdoid phages. If the DNA is terminally redundant, two possibilities are exonucleotic action at each terminus to form cohesive ends (followed by cohesive-end joining) and genetic recombination between homologous sequences.

15-43   Ligase, gyrase.

15-44   By cohesive-end joining between two DNA molecules.

15-45   L1, R1, R2, and R4.

15-46   No late mRNA is made. Therefore, there are no structural proteins, DNA concatemers are not cleaved, and the cells do not lyse.

15-47   Gene *N* can be mutant, because the deletion eliminates the ter-

mination site located before gene Q. The symbol *nin* stands for
N-independence.

15-48   T4 makes all of its own proteins and clearly does not need to use
         E. coli DNA as a template since it degrades the host DNA in the
         normal course of infection. λ apparently needs some host
         functions.

15-49   The products of genes *O*, *P*, and *dnaB* form a multisubunit protein
         complex. There is physical contact between the O and P proteins
         and between the P and DnaB proteins; the O and DnaB proteins are
         not in contact.

15-50   No, because T4 can be transcribed and the T4-encoded nucleases
         will destroy the λ DNA.

15-51   (a) Messenger RNA terminates at the first leftward terminator *tL1*
         of λ and *N* protein is needed to prevent termination. This
         conclusion is correct as long as the *trp* operon lacks a promoter.
         (b) Yes. If part of gene *N* is deleted, *tL1* must also be deleted
         and the mRNA from the main leftward promoter will extend through
         the *trp* operon.

15-52   The mutant *ti⁻* apparently fails to replicate only when other *ti⁺*
         phage are present; presumably the small burst size in the absence
         of any other phage is also due to a replication defect. It has
         functioning *O* and *P* genes. It is likely that it has a defective
         replication origin or *ori*. This functions weakly when no other
         phages are present but in a coinfection with a phage having a
         wild-type *ori*, apparently it cannot compete for the initiation
         system and replication fails to occur.

15-53   For the dimer, after the first DNA molecule is cut out and
         packaged, the remaining DNA has no *cos* site and hence cannot be
         packaged. Thus, only one particle can be formed from a dimeric
         circle. By the same reasoning, two particles are packaged from a
         trimeric circle.

15-54   Recombination produces a circular dimer in some of the infected
         cells.

15-55   (a) The E⁻ lysate contains active tails. These tails can attach to live *E. coli.* Filled heads from the *J*⁻ lysate attach to the preadsorbed tails. Following attachment, the λ *J*⁻ DNA can inject and the infection can proceed, producing *J*⁻ progeny.

(b) The genotype is *J*⁻ because this is the genotype of the DNA contained in the tailless heads.

15-56   (a) The *groE* mutation is recessive, since a λ phage carrying the *gro*⁺ allele can multiply in *E. coli groE*⁻.

(b) Phage B is missing the *int—red* region of λ but carries 4000 bases that include the wild-type allele of the *groE* gene.

(c) Phage C is phage B with an amber mutation in the *groE* gene.

(d) Protein X is the product of the *groE* gene.

(e) Protein Y is an *E. coli* protein whose gene maps near *groE*.

(f) Protein Z is a λ protein coded by a gene in the *int—red* region of the λ genome.

15-57   The (+) strand is the strand contained in the phage particle or any intracellular strand having the same base sequence. The (-) strand has the complementary base sequence.

(b) Yes.

(c) The DNA (-) strand is the coding strand, whereas the (+) RNA strand is the coding strand.

15-58   The point of this question is that if a phage codes for its own replicase or initiator for replication, then a successful infection can occur only if the nucleic acid in the phage particle is the coding strand. Also, if the replication origin were on a particular strand, then only that strand would be infectious.

(a) In ϕX174 the origin is on the (+) strand so the (-) strand is not infective.

(b) In the RNA phages the (+) strand encodes the replicase, so the (-) strand is not infective.

15-59   (a) It has overlapping genes.

(b) No, for it has so few genes that it surely must use host functions.

15-60   Only ϕX174 would make clear plaques because M13 does not lyse its host.

15-61   Bound to ribosomes.

Chapter 16

16-1 Temperate phage—a phage capable of lysogenization; lysogen—a bacterium carrying a complete set of genes of a temperate phage; defective lysogen—a bacterium carrying part of a set of genes of a temperate phage; integration—insertion of phage DNA into bacterial DNA; excision—removal of integrated phage DNA from a bacterial chromosome.

16-2 When a wild-type (cI^+) λ phage adsorbs to one bacterium, the lytic response ensues and many phage are produced that infect other bacteria. In time, the number of phage increases to the point that many bacteria are multiply infected. This favors the lysogenic response. Since the lysogens are immune to subsequent infection by λ, they grow in the plaque and produce a turbid growth of bacteria. A cI^- phage cannot establish repression and, therefore, always enters a lytic cycle; hence, an infected cell can never survive to form a colony. Phage T2 lacks a repressor and would not be expected to produce a turbid plaque.

16-3 Generally their repressors and repressor-binding sites (operators) must be the same, though it is sufficient if the repressor of each phage can recognize the operator of the other.

16-4 The phage has an amber mutation in one of the genes cI, cII, or $cIII$ and strain B (but not A) has an amber suppressor.

16-5 (a) No, because an operator mutation is cis-dominant.
(b) Yes, since the gene products are proteins and diffusible.
(c) Yes, but only by forming a double lysogen with a cI^+ phage.

16-6 Yes. In fact, if conditions are right for establishment of repression of one of the phage types, it will generally be right for the other also. Of course, if one of the phage types is not repressed, the infected cell will die.

16-7 (a) The repressor in the lysogen prevents transcription of the infecting phage DNA.
(b) Yes, because they would not be recognized by the repressor.

16-8 (b) Circularization is a physical process that does not need a phage gene product. Replication, however, requires prior transcription, which is inhibited by repression.

16-9   (a) The frequency with which a lysogen for a wild-type phage is
        spontaneously induced is relatively low, and the phage released
        form turbid plaques. However, if a mutation has occurred in the
        gene making the *cI* repressor, a lysogen containing this mutant
        (which will form a clear plaque) will always produce phage.
        (b) The gene *cI* makes repressor. Genes *cII* and *cIII* are needed in
        the wild-type state for the establishment of lysogeny, but not
        for its maintenance. Thus, a *cII⁻* or *cIII⁻* mutation in a prophage
        does not result in escape from repression.

16-10   The prophage attachment site is between genes *F* and *G*.

16-11

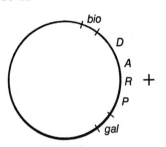 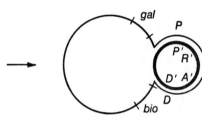

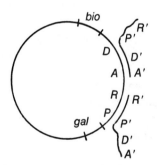

FIGURE A-16-11

16-12   When integrated into *E. coli*, the gene order is *gal cI P Q
        R A J bio* so *cI* will show the highest cotransduction frequency
        with *gal*.

16-13   (a) Repression or establishment of lysogeny.

(b) Complementation between two proteins was needed to establish repression.

16-14   Only N^- because it fails to make the cIII protein.

16-15   (a) λ *bio*, *POB'*; λ *gal*, *BOP'*.
        (b) λ *gal bio* has *BOB'*; λ *wild* has *POP'*;

16-16   Only λ*bio* x λ*bio* fails because neither phage has an *int* gene.

16-17   The plaque is turbid because repression can occur; only integration is defective.

16-18   A promoter-up mutation in the *xis* gene, possibly in *pI*; this would make production of integrase constitutive.

16-19   In the *cI*$^-$ infection, only mutant repressor is made, and this does not bind to DNA. In the *cI*$^-$ and *y*$^-$ infections, there is no complementation— *y*$^-$ is a site mutation preventing repressor synthesis by the DNA containing the mutation. The DNA of phage λ*imm434* is useful control because the cI repressor should not cause binding of this DNA. If one subtracts the binding activity of this phage from the binding activity of the *imm*λ phage, then one should know the binding activity of the *imm*λ repressor.
        (b) The *cro*-constitutive lysogen makes copious amounts of the Cro protein. This protein is seen to repress the *cI* function of the infecting *imm*λ phage. If a mixed infection is done, then one can see (last row of table) that the *imm434* phage supplies something to the *imm*λ phage that allows synthesis of the cI protein in an amount that prevents either synthesis or activity of the Cro protein. It is seen from the data that in this specific case, the Cro protein acts by suppressing the synthesis of the cII and cIII proteins by the *imm*λ phage.

16-20   (a) The prophage is transferred to a female lacking repressor; the operators are free of repressor and transcription begins.
        (b) Once this locus is transferred into the female, the female cells will die. Hence, by using the interrupted mating technique, one can determine at what time the numbers of any recombinant (for example, a locus transferred at an early time) decrease.

16-21   Decrease, because the *gal* and *bio* genes are moved apart by the prophage.

16-22   Using the integrase of $\lambda b2$ (λbio has no *int* gene), an Int-promoted recombinational event has occurred yielding a dimeric circle containing $\Delta OB'$ and POP', namely,

$$\Delta OP' \times POB' \rightarrow \Delta OB' + POP'$$

and the dimer has integrated using the newly generated POP' site.

16-23   (a) Prophage DNA can replicate, even though excision does not occur.
(b) The replication forks can leave the prophage and enter the bacterial DNA region.
(c) Excision must be fairly slow because often the replication forks leave the prophage before excision occurs.

16-24   If enough λ DNA molecules enter a bacterium, the intracellular repressor will be titrated.

16-25   (a) No, because in the absence of active repressor the *prm* promoter is inactive.
(b) Yes, because the renatured repressor activates *prm*.
(c) After ten minutes enough DNA replication has occurred without repressor synthesis that there is insufficient repressor to bind all of the operators. After five minutes repressor is still sufficient.
(d) Repression is re-established, but the Int and Xis proteins have been made, so prophage excision occurs. The excised prophage is repressed though and cannot replicate, so bacteria free of prophage DNA arise by successive cell division.

16-26   (2), because the Ter system excises and packages the segment of DNA between the *cos* sites of the two prophages.

16-27   Insertion has occurred at the right prophage *att* site (that is, at POB') to generate a dilysogen. The Ter system is active and cleaves at the two *cos* sites to produce a J^+P^+ recombinant. To prove this point, the prophage should be made A^+R^- and the infecting phage A^-R^+. All J^+P^+ phage will be A^+R^+.

16-28   The genotype will be $cI857P^+A^+$, because the only mode of phage production is Ter-mediated excision at the two *cos* sites. More phage particles are produced than in problem 16-26, because in problem 16–28 one prophage is P^+ and DNA replication can occur. A typical number of phage is twelve per cell.

16-29   (a) Infect individual cultures with λ*imm434int6* at 42°C. Because Ter excision of a tandem dilysogen can occur with the system of the infecting phage, each dilysogen will yield several *imm* λ phage. Such phage cannot be formed by this mechanism from a single lysogen. About one percent of the infected single lysogens will produce *imm* λ phage by genetic recombination between the infecting phage DNA and the prophage. If this is a problem, a recombination–deficient λ*imm434* (that is, λ *imm434red⁻*) can be used and the infection can be carried out at 32°C so that the prophage *red* gene will be inactive.
         (b) Strain B583 can be mated with an F^-sup^+ cell. Zygotic induction will occur and if the lysogen contains two prophage in tandem, phage will be produced.

16-30   (a) Three cistrons containing *(c1, c4)*, *(c2, c3)*, and *(c5)*.
         (b) Phage mutant *c5* must define a gene required to maintain lysogeny, since no lysogenic strains carrying *c5* are ever observed. The other mutants yield lysogenic strains at low frequency, so they must be able to maintain lysogeny once it has been established.
         (c) Clear mutants do not allow the temperate phage a choice between the lytic and lysogenic pathways of development, so no lysogenic bacteria ever grow up in the plaque centers.
         (d) A nonhomologous temperate phage may either lyse the cell, since it is not repressed for the unrelated phage, or it may undergo normal lysogenization, in which case the bacterial strain will become doubly lysogenic.

16-31   (a) Complementation.
         (b) The *dil* genes are required to set up lysogeny, that is, to establish repression, but are not needed to maintain the lysogenic state. Thus, the *dil* genes code for a positive control element in establishment of lysogeny.
         (c) The *tul⁻* lysogens are not stable, because there is no repressor. The *dil⁻* single lysogens can be isolated, since once repression is established with the help of the *dil* genes of the coinfecting *tul⁻* phage, the *dil* genes are dispensable. The situation is identical to that of λ, if the *tul* is the *cI* gene and the *dilA* and *dilB* genes correspond to the *cII* and *cIII* genes.

16-32   (a) To make the hybrid phage, grow P2 and 186 together in *E. coli*

K, and plate the progeny on *E. coli* C lysogenic for P2. This indicator selects against both parental types and allows the hybrid phage to plate.

(b) An *E. coli* F⁻*phe⁻ile⁻* lysogenic for *hP2 imm186* is constructed and mated with a nonlysogenic *E. coli* Hfr *phe⁺ile⁺* cell that is sensitive to the phage. Phe⁺ and Ile⁺ recombinants are selected. If the Phe⁺ recombinants lack the prophage, then the hybrid phage attaches near the *phe* gene. If the Ile⁺ recombinants usually lack the prophage, then the hybrid phage attaches near the *ile* gene. One should not use an Hfr*(hP2 imm186)* lysogen and a nonlysogenic F⁻ cell, because the phage would probably kill the recipient by zygotic induction.

(c) An *E. coli phe⁻ile⁻* strain lysogenic for *hP2 imm186* is transduced to Phe⁺ or Ile⁺ with P1 phage grown on prototrophic, nonlysogenic *E. coli.* If the Phe⁺ transductants tend to lack the prophage, then the hybrid phage attaches near the *phe* gene. If the Ile⁺ transductants tend to lack the prophage, then attachment is near the *ile* gene.

(d) The genes involved in prophage attachment are probably closely linked to the immunity gene, not the host range gene, since attachment specificity usually segregates with the immunity gene.

16-33   The vegetative map is *(a d c b)* and the prophage map is *(lac c b a d).* The two sets of data suggest that a linear, unique DNA molecule forms a circle which inserts by recombination between *c* and *d.* From known phages, the simplest guess for the DNA structure is one with complementary single-stranded ends. This might be tested by circle formation or end-to-end aggregation *in vitro,* as judged by sedimentation or electron microscopy.

16-34   Specialized only—integrates into the chromosome and either does not cause host DNA fragmentation during phage development or cannot package host DNA fragents. Generalized only—fragments host DNA, can package host DNA fragments, and probably does not integrate. Both—integrates, fragments host chromosome, and can package host fragments.

16-35   Integration does not occur during a lytic infection.

16-36   (1) Their excision system is perfect (unlikely). (2) When aberrant excision occurs, the fragment size is too large or too small to be packaged. (3) No known host genes are near enough to the prophage to be picked up.

16-37   This would be possible if they had a double-stranded DNA intermediate which inserted in the chromosome at a defined point.

16-38   The DNA in the transducing particles was replicated before infection. Also, either there is no replication of bacterial DNA after infection, or replicated DNA never gets into transducing particles.

16-39   The *tet-r* allele, and perhaps other RTF genes, are inserted within the prophage. The prophage excises normally, yielding a phage genome carrying *tet-r*; the DNA replicates, but the space inside the phage head is not sufficient to hold all P22 essential genes plus the inserted plasmid genes. Thus, the phage-like particles produced carry most, but not all, essential genes, and different phage particles lack different essential phage genes. This accounts for their cooperative growth behavior. The high frequency of Tet-r transduction is due to the fact that most particles will carry the *tet-r* allele. The frequency of Tet-r transduction cannot be 100 percent, as 90 percent of the infected cells are killed.

16-40   At each end of the prophage map essential genes are present that are absent in both types of transducing particles. Thus, lytic growth is only possible when a helper phage providing the missing functions is present. In single infection, transduction occurs by substitution because neither the right nor left prophage attachment sites can recombine with the bacterial attachment site. In mixed infection, either the transducing DNA and the helper phage DNA recombine or the attachment sites of the transducing DNA can recombine with a newly generated prophage attachment site.

Chapter 17

17-1   A circular DNA (and in one case, RNA) molecule (usually much smaller than the host chromosome) that replicates within a host cell doing the cell no particular harm and, in most environmental conditions, no good either.

17-2   Yes. The simplest plasmid would be a tiny DNA fragment carrying no information other than a copy of the replication origin of the host cell.

17-3 They are comparable.

17-4 (a) The clear colorless supernatant obtained from detergent treated cells from which the chromosome has been removed by sedimentation.
(b) No, only small ones.
(c) A cleared lysate is centrifuged to equilibrium in CsCl containing ethidium bromide.

17-5 The migration rate decreases with increasing molecular weight. Also the open circular form moves at about half the rate of a supercoil.

17-6 A plasmid to which plasmid-specific proteins, one of which has nuclease activity, are bound.

17-7 The plasmid has a temperature-sensitive mutation in some replication function.

17-8 (a) A conjugative plasmid is able to form a donor-recipient pair. A mobilizable plasmid carries genes needed for DNA transfer. A self-transmissible plasmid is both conjugative and mobilizable.
(b) All three.
(c) Conjugative.

17-9 (a) Conduction; no.
(b) No, it is not self-transmissible, and the chromosome is neither mobilizable nor conjugative.

17-10 (a) Rolling circle or looped rolling circle replication.
(b) A nick producing a 3'-OH terminus.
(c) Since an RNA primer is not needed to initiate rolling circle replication, a reasonable guess would be that transcription of an inducible gene is needed for transfer.

17-11 Synthesis in the donor provides the single strand that is copied. Synthesis in the recipient converts the transferred strand to double-stranded DNA.

17-12 $CaCl_2$ transformation using purified plasmid DNA.

17-13   (a) Grow the F^+ culture on agar containing acridine orange. Test colonies either for inability to transfer markers, ability to be a recipient or resistance to male-specific phages.
(b) F^- cells can act as a recipient, which immediately distinguishes them from donors. Growth of F^+ but not Hfr cells in media containing acridine orange leads to the formation of cells which act as recipients.

17-14   There are frequently two copies of an F' plasmid per cell.

17-15   Probably to bring donor and recipient cells in contact. It has been suggested that DNA passes through the sex pilus, but this idea has never been proved.

17-16   Of course, because the expression of F genes is the same in the integrated and nonintegrated states.

17-17   A gene in F must alter the cell wall so that the receptors on an F^- cell are absent in an F^+ cell. This view is confirmed by the existence of certain "female-specific" phages that adsorb to F^- but not F^+ cells.

17-18   Whereas F may in fact be transferred to an Hfr phenotypic female, incompatibility (namely, the F replication repressor) prevents F from being maintained in a colony that might develop from an Hfr cell that had received F. Incompatability is not a factor in detecting Hfr DNA because by recombination the transferred DNA becomes part of the chromosome of the F^+ cell.

17-19 - Either the homologous sequences are palindromes or the sequence in the chromosome at certain sites is duplicated with one copy inverted so as to generate a palindrome.

17-20   (a) Lac$^+$.
(b) Yes.
(c) F'lac has integrated into the chromosome. The chromosome is now replicating from the origin of F and hence does not need the dnaA gene product.
(d) No.

17-21   (a) Lac$^-$.
(b) F'(Ts)lac$^+$ has integrated into the bacterial chromosome.
(c) Integration has occurred inside a gal gene.

17-22   Either the replication repressor or the repressor binding site
        has been altered so that a higher repressor concentration is
        needed to shut off replication.

17-23   Five.

17-24   Early in the development of the colony a cell has divided to
        yield a daughter cell lacking $F'lac$.

17-25   (a) Since the transfer origin is in F and F DNA is transferred
        linearly, some F DNA is transferred first (just after $t = 0$) and
        the remainder is last (at $t = 100$).
        (b) Some of the cells in the Hfr population must have formed F'
        plasmids for transfer of a "late" gene to have occurred early.
        Thus, the genotype is $F'z^+/z^-str$-r.

17-26   This is the host modification-restriction phenomenon first ob-
        served with phages. In a K × B or a B × K mating the trans-
        ferred DNA is enzymatically destroyed in the recipient. In the
        few recipient cells in which restriction has not occurred the
        plasmid DNA gets the new host modification.

Chapter 18

18-1   A rearrangement of the base sequence of DNA.

18-2   RecA function and homology are needed only for homologous
        recombination.

18-3   (a) An organism which produces two types of progeny with respect
        to a single genetic characteristic, one having a (+) phenotype and
        the other having a (-) phenotype.
        (b) (1) A mismatched base pair and (2) two copies, one (+) and one
        (-), of the same gene. In case 2 heterozygosity appears when one
        allele is eliminated. The phenomenon is easily observed with a
        terminally redundant phage, like T4. For example, a single
        particle containing a (+) allele at one terminus and a (-) allele at
        the other terminus will produce progeny, half of which contain
        only the (+) allele and half of which contain the (-) allele.

18-4   Prophage integration and excision, formation of an F' plasmid
        from an Hfr, and formation of an Hfr cell from an F^+ cell.

18-5   (a) *RecX* is a general recombination gene and *recY* catalyzes site-specific recombination between genes *k* and *l*.
(b) All genes are equally spaced except *k* and *l* which are very near.

18-6   *Red⁻*; *red⁺* phage can acquire the *h* marker by recombination, but these mutants cannot.

18-7   Configurations 3 and 5.

18-8   Strand invasion occurred with the superhelix for a naturally occurring (negative) superhelix has long-lived, somewhat long, single-stranded regions, owing to transient unwinding. Invasion fails with nicked circles because, if the temperature were high enough to maintain large single-stranded regions, it would be too high for stable binding of a single-stranded fragment.

18-9   DNA-binding, proteolysis.

18-10 Single-stranded regions are generated during repair.

18-11 (a) Very late—so late in fact that often packaging occurs before the heterozygous intermediate can replicate or be mismatch-repaired.
(b) At least 10 percent. The number is a minimum, because some heterozygotes might not have been packaged.
(c) It is not very frequent but happens very early in the life cycle.
(d) When recombination occurs (which is infrequent), it happens repeatedly in the same cell and continues at late times.

18-12 Branch migration.

18-13 Two double-stranded DNA molecules connected by two nearby single strands.

18-14 (a) It must be possible to draw the observed configuration in a way that two circles of equal size are joined at one point.
(b) Adjust the DNA concentration to a very low value so that overlapping is very infrequent.

18-15 (a) No, the junction will drift by branch migration.
(b) The junction, moving back and forth by branch migration,

ultimately reaches an end of the X molecule and the two double-stranded units separate.

18-16 (i) Without the mutation. The principal observation is that most A^+R^+ recombinants have an intermediate density, because two DNA molecules having different density have recombined. Furthermore, recombinants have a broad range of density indicating that recombination can occur anywhere in the DNA. Since λ cannot package monomeric DNA, recombination must be used to generate a dimer circle. This conclusion is consistent with half of the phage of intermediate density being A^+R^+ recombinants; the other half are A^-R^-. The small peaks at high and low density have parental density and few A^+R^+ recombinants because these peaks contain particles produced by recombination between two phage having the same density and same genotype.
(ii) The presence of the mutation changes the distribution of recombinants—a large peak of A^+R^+ recombinants is just to the left of the light parental peak. If the mutation markedly stimulates recombination at a point 93 percent from the A end of λ, an excess of A^+R^+ recombinations whose density is 93 percent light and 7 percent heavy will result. Few recombinants are at the position of 93 percent heavy and 7 percent light for such recombinants would have both negative alleles and only ++ recombinants have been scored. Note also that more total phage particles are present: the higher recombination rate yields more dimers from which phage DNA can be packaged.

18-17 If 25 percent are a^+x^+, then presumably 25 percent are a^-x^-. Since pairing is random, for every pairing event between a phage and another of unlike genotype, a pairing event must occur between two particles of like genotype. Since 50 percent of the phage are recombinant, 50 percent must also be products of parents of like genotype. Therefore, all of the DNA participated in recombinational events.

Chapter 19

19-1 The total amount of DNA always increases when transposition occurs.

19-2 The transposon itself and the target sequence.

19-3   Direct repeat: ABCD....ABCD, in which the dots represent nonrepeated bases. Inverted repeat: ABCD....D' C' B' A' , in which X' is the complement to X.

19-4   One of the first-discovered bacterial transposons which lacked recognizable bacterial genes.

19-5   The fact that a replica of the transposon is found at a new position without loss of the original transposon.

19-6   The presence of a transcription stop signal in the transposon.

19-7   Heat denaturation followed by quick cooling and electron microscopic viewing might show a stem-and-loop structure. This would be suggestive but by no means would it be proof. Only base sequencing can answer the question for, if a transposon is present in the sequence, a target sequence must be duplicated. If a short duplication is seen, this is again only suggestive. If a duplication is not contained in the sequence, one can say with certainty that no transposon (of the types known) is present.

19-8   No.

19-9   (a) Integration of the plasmid into the chromosome by homologous recombination; reversion to temperature-insensitivity; transposition of the *amp* gene from the plasmid to the chromosome. (b) Integration (if homology-dependent).

19-10 You could start by forming heteroduplexes with a λ DNA molecule lacking the insertion. Each insertion will produce a single-stranded loop. If the loops are not the same size, the transposons must be different. If the loops are the same size, the two transposons are probably the same. A better test would be to move one of the transposons (A) to a nonhomologous DNA molecule. Then the DNA containing transposon A can be renatured with DNA containing transposon B. Only if the transposons are the same will heteroduplexes form. Radioactive RNA could also be obtained by *in vitro* transcription of the DNA containing A, and this RNA could be used in a hybridization experiment with the DNA containing B. Successful hybridization would indicate that the transposons are the same.

19-11 (a) The IS elements are inserted in two different orientations and there is a transcription stop sequence in only one orientation.

(b) Heteroduplexes between the DNA of the two phages would indicate that the transposons are the same or different.

19-12   Isolate two linear nonhomologous DNA molecules, each of which contain the transposon. These might be a plasmid DNA that had been cleaved to break the circle or a phage DNA. Mix the two DNA molecules, denature, and renature. Two types of molecules would result—perfectly renatured double strands obtained from the same DNA molecule and heteroduplexes having one double-stranded region, which is the transposon. The single-stranded termini are digested with exonucleases. The full-length double-stranded molecules and the transposons can be separated by centrifugation or, better, gel electrophoresis.

19-13   (a) The number of copies of the transposon in the new plasmid. These would be two copies if the fusion was mediated by a transposon.

(b) Form a hybrid with a nonhomologous molecule containing the transposon. If there are two copies of the transposon, a three-component structure consisting of one plasmid strand and two tester strands would be found. Base sequencing would of course also provide the information but that would be much more tedious. Restriction enzyme analysis (described in Chapter 20) would be easiest.

(c) Isolate the new plasmid and use the $CaCl_2$ technique to establish the plasmid in the Rec$^+$ cell. If there are two copies of the transposon, one component of the hybrid plasmid should occasionally be lost by a homologous Rec-mediated recombinational event.

19-14   (a) A high frequency of adjacent mutations suggests that a deletion including these genes has occurred. The deletion frequency is sufficiently high that one might reasonably suspect that a transposon is adjacent to the genes and that the phenomenon is an example of transposon-mediated deletion.

(b) Since many mutations result from SOS repair, which is absent in recA$^-$ cells, and since the transposition is independent of RecA function, study of the phenomenon in a recA$^-$ cell would be useful. If the deletion frequency were the same in Rec$^+$ and Rec$^-$ cells, a transposon would be implicated.

(c) Most mutagens would increase the frequency of chk$^-$ mutations with little or no effect on deletion frequency, if transposition were involved. There might be an effect if the site represented a mutational hot spot and the phenotype was a result of a pair of point mutations.

19-15 It seems possible that excision of a transposon would occur by means of RecA-mediated homologous recombination between duplicated target sequences.

19-16 (a) This technique is the standard penicillin-selection technique. Growth of rare cells lacking the transposon is inhibited by the tetracycline, so these cells are insensitive to penicillin as long as the tetracycline is present. The Tet-r cells are killed by the penicillin and only the Tet-s cells survive.
(b) Precise excision of the transposon; a deletion that either just removes part or all of the *tet* gene or the transposon plus adjacent DNA; a mutation in the *tet* gene.

19-17 (a) Because the size of genes is usually greater than the size of spacers and leaders, a great many mutations would be associated with having many transposons. Since each gene in a bacterial species serves a purpose, a large number of gene-inactivation events would surely weaken the cell and ultimately cause cell death. This probably is the reason that most transposons have been isolated from plasmids since in general a plasmid is not essential to its host cell.
(b) No, because the rapid accumulation of nonfunctional gene products, owing to repeated gene insertion, would lead to cell death.
(c) Only very slightly. Such a mutant might survive somewhat better in the laboratory because very rich culture media are used; these media provide so many nutrients that more mutations can be tolerated than in nature.

Chapter 20

20-1 Cloning is the act of inserting a particular gene or set of genes into a DNA whose replication can be controlled in order to produce many identical and easily recoverable copies of the gene or the set. Vector and vehicle have the identical mean-ing—namely, a DNA molecule into which a gene to be cloned can be inserted.

20-2 (a) An endonuclease that makes cuts in DNA at one particular base sequence.

(b) To destroy foreign DNA.

(c) Bacteria generally methylate one base in the sequence recognized by their own restriction enzymes and this methylation renders the sequence resistant to the enzyme.

(d) A type II enzyme makes cuts only within the sequence recognized. A type I enzyme makes cuts elsewhere.

(e) Each sequence has rotational (dyad) symmetry.

20-3   Flush or blunt ends are generated by single-strand breaks that are opposite one another. Cohesive ends are formed by single-strand breaks that are separated by a few nucleotides.

20-4   They must both recognize the same base sequence.

20-5   Both enzymes recognize the same base sequence. They might even be the same enzyme.

20-6   No. The two terminal fragments each have one blunt end.

20-7   Yes, as long as the fragments have different base sequence

20-8   The vehicle must contain at least one suitable restriction site into which foreign DNA can be inserted without destroying the ability of the plasmid to replicate. Ideally, the vehicle should also have some selectable markers and, even better, some marker that changes as a result of inserting DNA at the restriction site. If the latter is the case, cells carrying the recombinant plasmid can be quickly selected and identified.

20-9   By the CaCl$_2$ transformation technique.

20-10

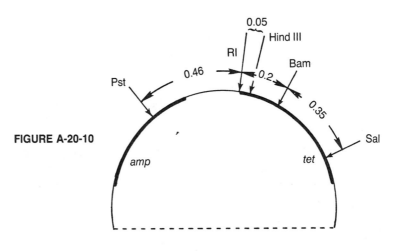

FIGURE A-20-10

20-11

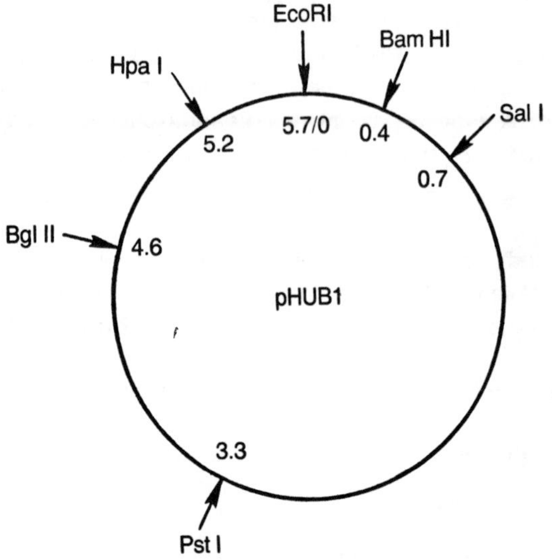

FIGURE A-20-11

20-12 Terminal transferase does not need a template. A poly(dA) strand can be added to one DNA molecule and a poly(dT) strand to another. The two molecules can be joined by hydrogen bonding between the poly(dA) and the poly(dT) tails. This is called homopolymer tail joining.

20-13 Blunt-ended fragments can be attached to one another by homopolymer tail joining with poly(dA) and poly(dT), or homopolymer tail joining coupled with adaptors having restriction sites.

20-14 The restriction enzyme is impure and contains at least one nuclease that is not site-specific.

20-15 Fragments often join together at room temperature by their cohesive ends. These short hydrogen-bonded sequences melt out at 65°C.

20-16 (a) Successful infection would be prevented by an alteration that would split a gene whose product is needed for replication and a rearrangement that would break the sequence in the replication origin.

(b) Yes, as long as it does not introduce the changes described in (a).

(c) The answer is the same as in (b).

20-17 At low concentrations circular and linear monomers are the principal products. At higher concentrations dimeric, trimeric, etc., products form. At early times the number of components per annealed unit is smaller than at later times.

20-18 The reaction has not been carried out long enough and the cuts at positions 35 and 83 were not made.

20-19 The terminal fragments are missing and a new fragment is present whose size is the sum of the sizes of the missing fragments. Thus, the DNA is circular.

20-20 (a) The order is 1.0-6.4-4.1. The reasoning is the following. First, in the linear DNA the 6.4-kb fragment can be linked to either of the other fragments, when digestion with Bam is partial; therefore, the 6.4-kb fragment must be in the middle. Second, in the circular DNA, where the cohesive ends are ligated together, the 1.0- and 4.1-kb fragments are joined, showing that they are at the ends.

(b) The P4 phage attachment site in the phage DNA lies in either the 1.0- or 4.1-kb Bam fragment and not in the 6.4-kb fragment, which is preserved after integration of prophage. The Bam sites in the host DNA that are nearest the prophage are placed to give host-phage fragments of 15.0 and 12.5 kb.

20-21 III-IV-I-V-II.

20-22 A plasmid dimer.

20-23 (a) Kanamycin.
(b) Kan-r Amp-r, Kan-r Amp-s.
(c) Kan-r Amp-s.

20-24 There are seven rrn "genes" with differing DNA sequences surrounding them. Since the Bam enzyme does not cut in rrn, there is one band for each rrn unit, and the size of the band is determined by the distance between the Bam sites nearest each rrn, on either side. Each rrn is cut by the Sal enzyme to give three pieces. When either labeled 16S or labeled 23S RNA is used as a probe, only two pieces from each rrn will get labeled. In each case, one of those labeled pieces will be very small and the other piece will have a size determined by the position of the nearest Sal site that lies outside of rrn. Thus, in the Sal columns the smallest pieces (seven of them) collect at the bottom, resulting in a thick dark band, indicating a greater than equimolar presence. The other seven pieces run to different positions on the gel.

20-25   The principle underlying the technique is that ^{32}P is found at intersections of overlapping fragments.

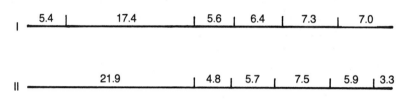

FIGURE A-20-25

20-26 $2f = 10^7/1.74 \times 10^{12} = 5.74 \times 10^{-6}$. Therefore, $N = \ln(0.01)/\ln(1 - 5.74 \times 10^{-6}) = 8 \times 10^5$ clones.

Chapter 21

21-1   Single or double-stranded DNA or RNA.

21-2   Single-stranded RNA viruses in which the virion RNA (i.e. RNA in the virus particle) is complementary to the viral mRNA.

21-3   (a) Enveloped viruses are encased in membranes.
       (b) Icosahedral and helical.

21-4   (a) Early and late refer to the time sequence of gene expression in viral development.
       (b) Viral DNA replication; turning various host cell functions off or on.
       (c) Synthesis of virion proteins.

21-5   Any kind that contains genetic material that is incapable of directing cellular enzymes to make mRNA. For example, (-)-strand

RNA viruses, like influenza virus; double stranded RNA viruses, like reovirus; double-stranded DNA viruses that multiply outside the nucleus, like vaccinia; and RNA viruses that replicate through a DNA intermediate, like retroviruses.

21-6 Retroviruses, which copy DNA from an RNA template, and vaccinia, which replicates its DNA in the cytoplasm where there are no DNA polymerases.

21-7 Capsid—the closed protein structure that isolates viral nucleic acid from the environment; protomer—the individual structural proteins of the capsid; capsomer—an organized aggregate of protomers; hexon—a capsomer consisting of six protomers; penton—a capsomer consisting of five protomers; virion—a complete virus particle.

21-8 Host cell outer membranes.

21-9 Segmented genome refers to a collection of viral genes that are distributed over two or more nucleic acid molecules; a heterocapsidic virus has a segmented genome in which different nucleic acid molecules are contained in distinct capsids; an associated virus is a defective virus that can multiply only when its host cell is also infected with a different virus ("helper virus"); a covirus is a plant virus having a segmented genome.

21-10 (a) Poly(A) tail, and except for poliovirus RNA, a cap.
(b) The (+) strand.

21-11 The density of the virus particle is less than that of 3 M potassium tartrate. This low density is probably a result of the virus containing lipid molecules or lipoprotein molecules. When these are removed by ether, the density increases.

21-12 (a) Release from surface vacuoles, lysis, and budding.
(b) The first two are used by naked viruses; enveloped viruses are released by budding.

21-13 Noninfective particles, experimental conditions that limit detectability, and a requirement for the participation of more than one virion for successful infection.

21-14 Double-stranded RNA viruses carry into the cell a replicase that copies double-stranded RNA and synthesizes a (+) strand; (-)-strand viruses carry into the cell a replicase that synthesizes mRNA from the (-)-strand template; (+)-strand RNA (other than retrovirus RNA) is translated directly; retrovirus RNA is used as a template for synthesis of DNA, which is then transcribed.

21-15   The (+)-strand viruses, such as poliovirus, synthesize a polyprotein that is cleaved to form individual proteins. Some (+)-strand viruses also make a (-) strand from which short (+) strands are made. The (-)-strand viruses of course first make a (+) strand. In some viruses this is cleaved; in others, a polyprotein is made, which is then cleaved; still others both process the (+) strand and cleave polyproteins.

21-16   Different cleavages of polyproteins; different cleavage patterns of mRNA, producing start codons in different reading frames; different splicing patterns of mRNA.

21-17   (a) The proteins labeled between two and three hours are degraded.
(b) A polyprotein is made at early times and it is degraded in at least two steps.
(c) A polyprotein is made that is first cleaved into three components. Since the proteins whose molecular weights are 25×10^3 and 15×10^3 are not present at late times, they must be unstable. The 5.6×10^5-protein is cleaved; since the sum of the molecular weights at late times is less than 5.6×10^5, either there are several copies of (at least) one cleavage product or one or more of the products is unstable.

21-18   Details are unknown, but either each replica remains associated with a replica of the core protein or, more likely, specific pre-core proteins bind to particular RNA molecules (base-sequence-specific binding) and these pre-core proteins aggregate specifically to form a core.

21-19   (a) Four—two copies of genomic (+)-strand RNA hydrogen-bonded together and two copies of host tRNA hydrogen-bonded to the genomic RNA.
(b) The tRNA molecules are primers and the genomic RNA molecules are templates for reverse transcriptase.

21-20   Integration is obligatory in the life cycle of a retrovirus.

21-21   Entry of viral proteins into the infected cell, processing of mRNA, production of polyproteins, conversion of RNA to DNA.

21-22   Supercoiled DNA. Occasionally, if the DNA is not isolated carefully, open circles and even linear molecules can be found. If the DNA is not subjected to procedures that remove proteins, all histones other than H1 are associated with the DNA, which has a nucleosome-like form.

21-23   A gene product of SV40 and polyoma viruses, originally detected in tumor cells, which is necessary for initiation of viral DNA replication. The component first termed T antigen (with polyoma) contained T, t, and middle-t antigens.

21-24   (a) A procedure for isolating viral DNA by precipitation of high-molecular-weight DNA, leaving viral DNA, which has a low molecular weight, in the supernatant.
(b) There is not a good one, though the amount of DNA isolated from a particular number of cells is quite reproducible.

21-25   Differential splicing allows different reading frames to be used and removes particular start and stop codons. Some proteins are translated in different reading frames and others are totally contained in the sequence of a larger protein. No proteins are cleaved from polyproteins.

21-26   The viral DNA must replicate in the nucleus, which is the location of the polymerase.

21-27   The 5' termini of the two single strands of adenovirus double-stranded DNA are covalently linked to a protein molecule that dimerizes.

21-28   The 5'-terminal protein first becomes covalently linked to the 5'-α-P of dCTP. The 3'-OH group of this molecule is sufficient for priming.

21-29   (a) Permissive—progeny virus are produced; nonpermissive—viral nucleic acid enters the host cell but progeny virus are not produced.
(b) Productive infection—same as permissive; transformation—virus particles are not produced and the host cells acquire

particular growth characteristics not present in uninfected cells.

21-30   Cells which, in nature, grow in the form of tissues grow in cell culture until the cells are in contact with one another. At this point they have formed a monolayer and stop growing. Tumor cells continue to grow and form disorganized, piled-up masses of cells.

It is now known that it is cell density and not physical cell-to-cell contact that is relevant.

21-31   Transformed cells do not show density-dependent growth or anchorage dependence, whereas normal cells do. They require less serum for growth and are more easily agglutinated by lectins than normal cells. A transformed cell is presumably a prototype for a tumor cell.

21-32   (a) Cells are infected with Sendai virus that has been inactivated by ultraviolet light. Infected cells in contact frequently fuse.
(b) The maximum is 52. The minimum would be 23 or 31, depending on which cell line was donating the extra chromosomes.

21-33   By cell fusion with permissive cells. If this results in induction of formation of complete virus, the entire viral nucleic acid must be present, unless the virus happens to possess nonessential genes.

21-34   Tumor viruses do not have repressors. Integration of viral DNA probably does not occur at uniquely genetically defined sites. Integration of viral DNA is not catalyzed by a virus-encoded enzyme. Transformation occurs primarily in nonpermissive cells, whereas whether a phage undergoes a lytic or a lysogenic response depends on the multiplicity of infection and the growth state of the bacteria rather than on the genetic characteristics of the bacterium. Some viruses integrate a DNA fragment and still produce transformation, whereas prophages are invariably complete phage DNA molecules.

21-35   A mutant lacking some function that kills the host cell might transform. Interestingly, replication-defective SV40 viruses can transform permissive monkey cells.

Chapter 22

22-1   Repetitive sequence—a base sequence that is repeated, usually many times, in the DNA of an organism; unique sequence—a base sequence for which there is only one copy per haploid set of chromosomes; satellite DNA—a repetitive sequence whose base composition is sufficiently different from the average base composition of the organism that the DNA can be separated on a CsCl density gradient from the bulk of the cellular DNA.

22-2   (a) The DNA is about 30 percent unique, about 60 percent redundant, and the remaining 10 percent is highly redundant satellite DNA.
(b) The three classes have Cot midpoints of about 10^3, 10^{-1}, and 10^{-3}, respectively. Thus, the number of copies per genome is $10^3/10^3 = 1$ for the unique DNA (by assumption), $10^3/10^{-1} = 10^4$ for the redundant DNA, and $10^3/10^{-3} = 10^6$ for the satellite DNA.

22-3   (a) Eukaryotic genes are not generally organized into polycistronic operons since, without complex splicing events, only one protein molecule could be translated from the primary transcript.
(b) An operon refers to a set of coordinately regulated genes that are encoded in one or two polycistronic mRNA molecules. If there is a regulatory gene (for example, a repressor), it is usually encoded in a separate monocistronic RNA molecule. A gene family is a collection of genes, each of which yields a distinct mRNA molecule, that encode molecules of similar or related function. The genes may or may not be clustered and they are rarely adjacent, but they are somehow regulated by similar or related signals.

22-4   There are many instances in which several distinct protein molecules, having like or identical function, are prevalent at various stages of development of an organism. The differences between the proteins are often only a few amino acids. The different forms of the protein are encoded in distinct genes, each of which is active at a particular stage of development. A collection of such genes comprises a developmentally regulated gene family.

22-5   (a) No. (b) No. (c) No. (d) Between the genes.

22-6 (a) An adult cell is considered to be totipotent if it retains the ability to direct embryonic development when transplanted into an embryo. It is thought that such a cell has not lost any embryonic genes; however, since totipotency is experimentally defined by a transplantion experiment, it is possible that some genes present in the embryo, but not necessary for embryonic development, may be lost.

(b) A mechanism might exist for amplifying genes yet there may be no mechanism for gene removal; a repressor-like protein, which is synthesized only once in a cell, may be removed and irreversibly activated; a gene rearrangement may occur; a positive regulator may be activated or synthesized yet there may be no way to deactivate it or to prevent its synthesis; there may be a change in chromosome or chromatin structure (chromatin activation).

22-7 (a) Since the ribosomal genes replicate extrachromosomally, some type of recombinational event is needed to excise the rDNA.

(b) Nucleases to catalyze recombinational excision. These might be endo- and exonucleases having additional functions or there might be a site-specific enzyme like the phage λ integrase. Since rolling circle replication is utilized in the amplification, there must be an endonuclease (site-specific) that makes the initiating nick for this mode of replication.

(c) The excised rDNA units fail to replicate and are diluted out as cell division proceeds.

(d) The signal is probably one that prevents initiation of rolling circle replication. Fertilization might cause synthesis of an inhibitor of the initiating endonuclease referred to in (b), of a repressor protein that makes the replication origin inaccessible, or an exonuclease that attacks the linear part of the rolling circle until the entire structure is degraded.

22-8 Limited time—gene amplification; long time—increased lifetime of mRNA with or without gene amplification.

22-9 The gene encoding the protein could have a unique promoter. An effector molecule could cause synthesis of two new proteins, an inhibitor of the cellular RNA polymerase, and a new RNA polymerase that could recognize only that promoter. You may have other ideas.

22-10 An embryonic gene contains a large number of different base sequences that constitute the coding sequences for all antibodies. These sequences are contiguous in the DNA. In the course of development a genetic recombinational event removes large blocks of DNA that include many adjacent sequences. There are many blocks that can be removed, so that many different coding sequences can remain after this recombinational event occurs. One event occurs in a particular cell leaving that cell with a unique coding sequence that enables it to make a particular antibody.

22-11   First, a cell has engaged in a recombinational event that enables it to make a particular antibody, as described in problem 22-10. In a way that is not understood, the cell is also programmed to multiply on continuous (or subsequent) exposure to the antigen. Thus, a clone of cells, which make a single antibody, forms. This response, which is called clonal selection, is an example of cell amplification.

22-12   (a) Increase the number of J genes.
        (b) $150 \times 12 \times 3 \times 5000 = 2.7 \times 10^6$.

22-13   A cell can respond to a particular hormone only if the cell has a receptor binding site for that hormone on the cell membrane.

22-14   In gene amplification the number of copies of a gene is increased; therefore, more mRNA can be made per unit time. In translational amplification the lifetime of the mRNA is enhanced, which enables a great deal of mRNA to be present at a particular time.

22-15   There is no way to predict this because mRNA abundance is not usually related to copy number.

22-16   (a) The cell must not require continuous synthesis of any essential protein.
        (b) Globin is synthesized in reticulocytes, which have no nucleus.

22-17   Binding of translation factors could be different in the two mRNA molecules. One type of mRNA could be less stable.

22-18   The specificity lies in the chromatin not in the extract and some inhibitor of oocyte transcription is washed from the chromatin by the 0.6 M NaCl.

22-19   The hormone is apparently needed to initiate intron excision.

22-20   E somehow regulates excision of a single intron that contains an AUG codon that is nearest to the 5' terminus of the RNA. There is also an in-phase stop codon that may or may not be in the intron. Once the intron is excised, there is another AUG codon that initiates synthesis of Q.

22-21   (a) Digest chromatin from each cell type with DNase I to a limited extent of digestion (10 to 20 percent of the DNA rendered acid-soluble) and then hybridize with labeled c-DNA or labeled mRNA (if available). If the gene is in active chromatin, it will be destroyed by the enzyme more rapidly than the bulk DNA and not be available for hybridization.
        (b) Digest with micrococcal nuclease and then hybridize with labeled c-DNA or mRNA. If the gene is in a nuclease, it will be resistant to the enzyme.

22-22   (a) Measure the number of genes by Cot analysis and the abundance of the mRNA by a hybridization procedure. You would be interested in enhanced transcription because you have linked the gene to a different promoter.
        (b) Grow the cells in the absence of methotrexate and see if the extra copies are diluted out by cell division (one sign of nonreplicating extrachromosmal DNA). Alternatively, use the Southern transfer procedure to seek large pieces of DNA of discrete size containing both plasmid and cellular sequences.
        (c) Gene amplification usually occurs as amplification of a large segment of DNA so that one would expect both genes to be amplified. Take the *xgp* portion of the plasmid and use as a probe in a Southern transfer experiment. This procedure is useful for amplifying any gene by linking it to the *dfr* gene.

22-23   (a) About 50 percent of the sequences synthesized in the late oocyte are not present in the gastrula. All of the sequences synthesized in the gastrula are present in the late oocyte.
        (b) None. These experiments use only cytoplasmic RNA as a competitor.
        (c) Use nuclear RNA rather than cytoplasmic RNA in an otherwise identical set of experiments.

22-24   (a) The promoters for both transcription units probably have a common sequence acted on by either a positive or negative regulator.

(b) Both primary transcripts have a common sequence acted on by an element that prevents some stage of processing. (c) Both processed mRNA molecules have a common sequence involved in ribosome binding. An effector may remove a protein bound to this sequence or it may denature a double-stranded region containing the ribosome binding site.

22-25   (a) The *B* gene may function because it lacks introns. It is unlikely that the *A* gene would function unless it happens to lack introns.
(b) It seems as if the gene cannot function without its introns. There is some precedent for this in the globin system—globin mRNA is less stable in the cytoplasm if it is formed by direct transcription of the coding sequence of DNA than if it is spliced from intron-containing RNA.

22-26   No. Note the single-stranded poly(A) tails projecting from the junctions. These are always at the 3'-OH terminus of the RNA yet they are not at the same ends of each loop. Thus, the RNA molecule of the central gene is transcribed from the strand that is complimentary to the coding strand for the outer genes.

Appendix A

The "Universal" Genetic Code*

First position (toward 5' end of strand)	Second position				Third position (toward 3' end of strand)
	U	C	A	G	
U	Phe	Ser	Tyr	Cys	U
	Phe	Ser	Tyr	Cys	C
	Leu	Ser	Stop	Stop	A
	Leu	Ser	Stop	Trp	G
C	Leu	Pro	His	Arg	U
	Leu	Pro	His	Arg	C
	Leu	Pro	Gln	Arg	A
	Leu	Pro	Gln	Arg	G
A	Ile	Thr	Asn	Ser	U
	Ile	Thr	Asn	Ser	C
	Ile	Thr	Lys	Arg	A
	Met[†]	Thr	Lys	Arg	G
G	Val	Ala	Asp	Gly	U
	Val	Ala	Asp	Gly	C
	Val	Ala	Glu	Gly	A
	Val	Ala	Glu	Gly	G

*The "universal code" does not apply to mitochondria.
[†]AUG is a codon that also signals "start," though not in all cases.

Appendix B

Genetic Map of *E. coli,* Showing Relative Positions of Selected Markers

Abbreviations (clockwise from top): *thr,* threonine; *leu,* leucine; *lac,* lactose; *lon,* a repair gene; *gal,* galactose; *att* λ, prophage attachment site for phage λ ; *bio,* biotin; *trp,* tryptophan; *his,* histidine; *pur,* purine; *recA,* a recombination gene; *ser,* serine; *thy,* thymine; *fda,* aldolase; *str,* streptomycin; *metE,* a gene in the methionine biosynthetic pathway; *argC,* a gene in the arginine biosynthetic pathway; *malB,* a gene needed for maltose breakdown; *uvrA,* gene for excision repair.

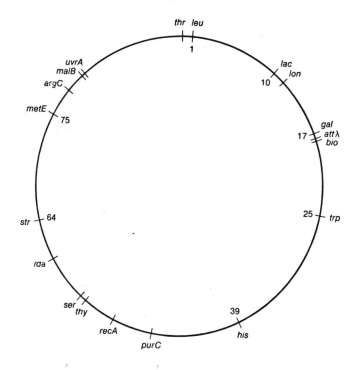

Appendix C

Genetic and Physical Map of *E. coli* Phage λ

The heavy line represents the DNA molecule present in the phage head. Genes are indicated in the correct position on the DNA. An expanded view of the *att*–*cI* region is shown. Abbreviations: *m*, *m'*, the single-stranded cohesive ends which, when joined, form the *cos* site; *A*, the gene for producing *m* and *m'* from a *cos* site: *E*, the gene for the major head protein; *K* and *J*, genes for tail proteins; *b2*, a nonessential region which is sometimes deleted; *att*, the prophage attachment site, also refered to as *POP'*; *int*, integrase; *xis*, excisionase; *red*, a locus consisting of two genes for homologous genetic recombination; *gam*, a gene responsible for inhibiting the *E. coli* RecBC protein; *cIII*, a gene involved in the synthesis of both integrase and the repressor; *tL1*, the first leftward terminator from *pL*; *N*, a gene whose product allows *tL1* to be ignored; *pL*, the leftward promoter; *oL*, the leftward operator; *cI*, the repressor; *O* and *P*, two genes needed for initiation and continuation of DNA synthesis; *Q*, a gene required to initiate transcription of the *SRAEKJ* region; *S* and *R*, genes involved in lysis.

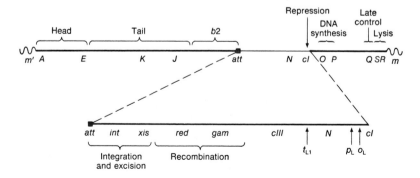

Appendix D

The Poisson Distribution

If a large number N of balls are placed randomly into a large number r of boxes, and if the number of balls exceeds the number of boxes, each box will not contain the same number of balls. If r is much greater than N, many boxes will be empty, most boxes will contain one ball, but some will contain more than one ball.

The Poisson distribution enables one to calculate the fraction $P(n)$ of the boxes containing 0, 1, 2, ... n balls in terms of $N/r = a$, the average number of balls per box. The distribution is generally written

$$P(n) = \frac{a^n e^{-a}}{n!} \qquad (1)$$

Thus, if the average number of balls per box is 3, then $P(0) = 0.0497$, $P(1) = 0.15$, $P(2) = 0.22$, and so forth. An importance consequence of this distribution is that

$$P(n) = e^{-a}. \qquad (2)$$

This is the so-called "Poisson zero" and enables one to determine a, and hence N.

The Poisson distribution applies to many biological situations. An example is the multiplicity of infection of a phage population infecting a bacterial population. If the fraction of bacteria surviving the infection (that is, the uninfected cells) is measured, this value is the Poisson zero and a in equation 2 is the multiplicity of infection. Another example is the determination of the rate of production of damage to a population of molecules. If a DNA sample is exposed to various doses of x rays, molecules are broken and the amount of breakage increases with dose. A graph can be made of the fraction of the molecules that remain unbroken as a function of x-ray dose. The dose that yields a fraction unbroken of $1/e$ is the dose at which there is, on the average, one break per molecule. Thus, the rate of breakage is one break per unit dose per unit molecular weight.

Index

by Problem Numbers